The Evolution
of
Differentiation

The Evolution
of
Differentiation

WILLIAM S. BULLOUGH

Birkbeck College, University of London
London, England

1967

ACADEMIC PRESS

LONDON AND NEW YORK

ACADEMIC PRESS INC. (LONDON) LTD
Berkeley Square House
Berkeley Square
London, W.1

U.S. Edition published by
ACADEMIC PRESS INC.
111 Fifth Avenue
New York, New York 10003

Library of Congress Catalog Card Number: 67–31046

PRINTED IN GREAT BRITAIN BY ROBERT MACLEHOSE AND CO. LTD
THE UNIVERSITY PRESS, GLASGOW

Preface

This is a book that cannot be written adequately. No one person can compass the fields of microbiology and molecular chemistry, of genetics in both unicellular and multicellular organisms, of embryology in all its many aspects, of tissue homeostasis including endocrinology, and of the two great problems of ageing and cancer. Indeed, so voluminous is the literature, and so rapidly is it being produced, that it is well-nigh impossible for one person to compass any one of these fields. Yet in any consideration of that central biological problem of gene control, or differentiation, the evidence from all these fields is highly relevant, and it is clear that no active progress can be made until some reasonable framework has been erected within which the diversity of facts can be fitted and on which new experiments can be based. This book is an attempt to sketch the outlines of such a framework from what is now known or suspected of the various aspects of differentiation in all types of organisms. No apologies are offered for any weaknesses in detail; it is the main argument that matters.

This argument is not entirely new. It has been slowly emerging for many years, and the greatest single contribution has come from the microbiologists. Above all it is the so-called Jacob–Monod theory of gene control that has enabled the pieces of information from a wide variety of sources, from protozoan differentiation to mammalian carcinogenesis, to be assembled into a meaningful pattern. However this pattern is still very incomplete, and in attempting to describe it here it has often been necessary to speculate well beyond the established facts. Again no apology is offered. Many of these speculations may prove erroneous, but in provoking argument and experiment they may still serve a useful purpose.

The ideas put forward in this book have been developed over many years during which many debts have been incurred to many people, but unfair as it may be, it is not possible to thank them all individually. However, three people have been imposed upon and have suffered daily

during the whole period of preparation of the manuscript. These are
Dr Edna B. Laurence, Mrs J. U. R. Deol and Miss D. Speller, and although it may seem commonplace to say that without their help this
book could never have been written, in this instance it is strictly true.

July, 1967   *William S. Bullough*

Contents

CHAPTER 1

Introduction

It is generally agreed that one of the central problems in biology today is that posed by the process of cellular differentiation, while one of the central problems in medicine is that posed by the distortion or collapse of this cellular differentiation, which may lead in embryonic life to developmental abnormalities and in post-embryonic life to cancer. As a result of decades of frontal attack there now exists a vast literature on these biological and medical problems, but in both cases it seems that successful solutions are more likely to come from flank attacks such as are now being made from a number of directions.

Recent years have seen a great increase in our knowledge of cell biology, especially at the macromolecular level. Most important has been the clear demonstration that the well-known physical similarities between cells, the common possession of cytoplasm and nucleus both with typical organelles, are simply an expression of the remarkable uniformity of the basic cell mechanisms. In particular, all cells store information and use information in the same way.

It is now believed that most of the essential information about an organism is written along the length of its chromosomes in a code composed of triplets of four nucleotide bases and that in this form it is preserved from generation to generation. In all cells the directions in this code are implemented in the same way. The message is transcribed on to messenger RNA, is then interpreted on the polyribosomes (which are themselves made under DNA instructions), and ultimately is translated into the amino-acid sequences of the various enzymes (with the assistance of transfer RNA molecules which are also synthesized under DNA control).

In the code one nucleotide triplet specifies one amino-acid; one group of triplets, which is a gene, specifies one amino-acid sequence, which is an enzyme; and one functional group of genes, which is an operon, specifies a sequence of enzymes, which together control one particular synthesis. Thus in the language of the DNA code, a triplet can be regarded as a letter, a gene as a word, and an operon as a sentence. The alphabet of the language has been compiled by Stretton (1965).

It is obviously of interest to discover whether the same genetic language is used by all species of animals and plants, and the limited evidence so far available suggests that in essence this may be so (see, for instance, Gros, 1964; Lanni, 1964). However, it does appear that certain local dialects may exist; for instance, some amino-acids, such as leucine, are already known to be specified by more than one triplet, and it seems probable that one organism may tend to use one and another organism another. There are 64 possible triplets which are used to code for only some 20 amino-acids; some of the excess triplets are evidently alternatives while others appear to be "nonsense" triplets. It is, however, by no means certain that the "nonsense" triplets are lacking in function, and there is evidence that they may be used, for instance, to signify the end of a gene word or of an operon sentence (Stretton, 1965).

It must also be stressed that not only may the DNA issue essentially the same basic instructions in the cells of all species, but that the DNA of all species may also respond in essentially the same way to the presence of particular enzymes. Thus, for instance, RNA polymerase from the bacterium *Escherichia coli* will readily support DNA-dependent RNA synthesis in the cells of animals such as ducks and rats and of plants such as peas and potatoes (Bonner, 1965).

So the coding of the instructions and the mechanism of their interpretation appears to have been influenced only slightly during the whole course of evolution since the time of appearance of the first modern type of cell. The wide diversity of cell form and function shown today by unicellular and multicellular plants and animals can be regarded merely as the result of a myriad variations on certain old-established themes. These variations are the outcome of at least two interrelated processes: first, the evolutionary selection of new and radical mutations which has resulted again and again in the creation of a novel genome; and second, the evolutionary elaboration of that mechanism whereby each cell is able to select the instructions it will follow from among all the instructions that are coded on its chromosomes. Thus the epidermal cells of a mammal selectively respond to instructions to synthesize keratin, while the erythroblasts, which possess exactly the same genes, respond instead to instructions to produce haemoglobin.

It is this choice of programme, whereby certain genes are activated while others are repressed, which is the essence of differentiation.

The term differentiation was originally used to describe the process of origin of the tissue cells of the higher organisms. Now, however, it is becoming clearer that differentiation must be considered in a much wider context, and that it may be defined as that process whereby any cell, whether of a unicellular or a multicellular organism, is able to respond to the demands made on it by selecting between the alternative potentialities of its total gene complement. Defined in this way the phenomenon ranges from the readily reversible state of differentiation seen in a micro-organism in the presence of a particular food material, through the potentially reversible state of differentiation seen in the tissue cells of higher green plants, to the apparently permanent state of differentiation seen in the tissue cells of mammals.

Micro-organisms, such as *E. coli*, are cells whose main armament in the face of an ever-changing environment is their ability to select among the extensive possibilities of their genomes so as to synthesize those enzymes, and only those enzymes, that are needed at any moment. Monod and Jacob (1961) have emphasized that each pattern of enzyme synthesis should be regarded as the outcome of a transient state of differentiation, and that each particular expression of the genetic possibilities is evoked by some specific chemical messenger (see also Jacob, 1964; Jacob and Monod, 1963; Monod *et al.*, 1963).

In the higher plants the differentiated state of the cells is normally permanently maintained, although it has long been known that in many species small pieces of stems or of roots, or even small pieces of leaves, can give rise to whole plants. Recently it has been shown that a single isolated differentiated cell, for instance of a carrot leaf, when appropriately stimulated with plant hormones, can be induced to give rise to a complete new plant (see Steward, 1963; Steward *et al.*, 1963). This observation is particularly important in indicating that when, in a plant tissue cell, gene expression is severely limited, there is no destruction of, or even damage to, the repressed and unused genes. The genome remains at least potentially totipotent.

In the higher animals, such as mammals, it has not yet proved possible to reopen the closed regions of the genome in a differentiated cell, and although a technique may perhaps ultimately be found, it appears that the unused genes are particularly firmly repressed. However, it is not generally believed that these genes are thereby necessarily damaged or destroyed. Also, it must be emphasized that in most fully-formed mammalian tissues some limited selection of gene expression usually remains possible between a few groups of genes which are not firmly repressed (see p. 102). Thus at any given moment a tissue cell may contain active genes, lightly repressed genes and firmly repressed genes. The lightly repressed genes are almost as readily activated by specific chemical messengers as are the genes of the micro-organisms. Faced with this situation in mammalian tissue cells, Foulds (1964) has spoken of the total genome (the entire genetic information within the cell), the facultative genome (the genes that remain available for activation and use as circumstances require), and the effective genome (the genes that are functional at any given moment).

In general the evolution of all types of multicellular organisms has been accompanied by an increasing stability of cellular differentiation, and indeed it is obvious that such stability in the tissues of the higher organisms is as essential to their survival as is lability of differentiation to the survival of the micro-organisms. However, this change in evolutionary emphasis may have involved little or no change in principle, and throughout the whole range of living organisms it is commonly reported that gene activation or repression, whether temporary or permanent, always seems to occur in response to the trigger-like actions of certain chemical messengers, which extend from the effector substances of the bacteria to the hormones of the higher plants and animals.

The aim of this book is to survey the many aspects of differentiation that are seen today, and in particular, because of the evident universality of expression of the genetic code, to enquire whether any similar universality exists in the methods of gene repression and activation. The information reviewed ranges from the chemical control of the genetic mechanisms of unicellular organisms during their various phases of activity, to the chemical control of differentiation in multicellular organisms both during embryonic development and in the adult state. Finally some consideration is also given to that collapse of differentiation which leads to cancer.

CHAPTER 2

The Simplest Living Organisms: Bacteria

It is generally believed that during the development of those chemical organizations that ultimately became recognizable as "life", there must have been a long period of chemical evolution leading to the creation of a variety of dynamic, self-perpetuating, chemical equilibria, which were somehow stabilized as for instance in the interstices of fine clays (Oparin, 1957, 1962; Bernal, 1951; Fox, 1965). It is possible that the more successful of these equilibria were formed by the fusion, or symbiosis, of less efficient equilibria which possessed complementary metabolic aspects. At some point in this evolutionary process a major step forward evidently occurred when for the first time some local chemical unit became enclosed in a membrane, and it may be that this first cell was then able to drive into extinction all the remaining non-cellular chemical equilibria. It is even possible that this first cell already possessed the mechanism whereby all activity was controlled by DNA, effected by RNA, and mediated by enzymes, the actions of which led to the production of more DNA and RNA and of more cells.

It is reasonable to suppose that, when the era of chemical evolution gave place to an era of cellular evolution, there may have arisen within the cells a considerable diversity in the arrangements of the biochemical reactions, some visible perhaps as organelles. If this is so then only the most efficient of these have survived to be represented today by two highly successful cell types. The first of these is represented by the cells of the blue-green algae and of the bacteria (from which the viruses have evidently been derived by simplification), and it is typified by the lack of a nuclear membrane, by chromosomes composed entirely or almost entirely of DNA, and by a simple amitotic form of cell division. The second surviving cell type is present, sometimes with considerable modifications, in all other living organisms, and it is typified by the possession of a discrete nucleus and cytoplasm each with a number of typical organelles, by chromosomes in which the DNA thread is strengthened by combination with proteins (histones and occasionally protamines), and by a complex method of cell division by mitosis.

THE ANTIQUITY OF THE BACTERIA

In searching for the basic mechanisms of gene control and thus for the possible origins of differentiation, it is obviously important to explore the cellular mechanisms of the most primitive organisms that have survived to the present day, and it seems highly probable that these are the bacteria and the blue-green algae (Glaessner, 1962; Echlin and Morris, 1965).

The fossil record allows the evolutionary origins of the major plant and animal phyla to be traced in moderate, if discontinuous, detail as far back as the Cambrian, which was roughly 500 million years ago. The earlier fossil record is remarkably poor, but in recent years the range of available information has been extended back for at least 3000 million years. At the moment the earliest evidence of life comes from a Swaziland series of sedimentary rocks, which are probably older than 3000 million years (Barghoorn and Schopf, 1966); similar evidence also comes from the Gunflint chert (or flint) from the northern shore of Lake Superior (Barghoorn and Tyler, 1965), a deposit which is about 2000 million years old. Both these series of rocks contain an abundance of structurally preserved fossils, the most prominent of which resemble the modern blue-green algae while the rest resemble modern bacteria. Chemical analysis of the chert also indicates the presence of organic matter which was probably produced by photosynthetic organisms.

Thus the bacteria and the blue-green algae not only possess the simplest known cell structure but also are by far the earliest organisms in the fossil record. They are clearly the organisms of choice in any search for the most primitive surviving manifestations of gene control and of differentiation, and it is fortunate that with the introduction of micro-organisms into biochemical genetics a considerable amount of information on the basic cell mechanisms of the bacteria, as well as of viruses, is now available.

CELL MAINTENANCE AND GROWTH

The life cycle of a bacterium commonly includes at least four major phases of activity: a period of cell maintenance and growth, a period of cell division, occasionally a period of sporulation, and perhaps less commonly a period of conjugation during which genetic information is passed from individual to individual. There is good evidence that each of these phases is controlled from the gene level and that the transition from one phase to another involves a change in the type of syntheses being conducted within the cell, which is itself the outcome of a change in the pattern of gene repression and activation.

The survival and growth of any bacterium naturally depends on the availability of an adequate supply of suitable food which can be utilized for the synthesis of all those proteins that are needed for all aspects of metabolism. For the utilization of the available foods appropriate enzymes must be synthesized, and for the synthesis of essential cellular constituents other appropriate enzymes must be synthesized. All these various processes work in unison, and it is obvious that the system as a whole is able to regulate itself by means of specific feedback circuits. In the words of Changeux (1965) "we can think of the cell as a completely automatic chemical factory. . . . Regulating the production lines are control circuits that themselves require very little energy. Typically they consist of small, mobile molecules that act as 'signals' and large molecules that act as 'receptors' and translate the signals into biological activity".

These control circuits are of two main kinds. The first depends on a system whereby the adequate accumulation of an end-product automatically inhibits the activity of the first enzyme of the metabolic sequence leading to its synthesis (see Changeux, 1965; Kornberg, 1965). The classical example is provided by the bacterium *Escherichia coli*, in which an excess of the amino-acid L-isoleucine immediately inhibits the activity of L-threonine deaminase, the first enzyme in the synthetic pathway (Fig. 1). This inhibition may be due to a distortion produced in the enzyme molecule when it combines with the end-product.

The second kind of control circuit, which is more relevant to the present argument, depends on a system of gene activation and repression, the elucidation of which is particularly associated with the work of Jacob and Monod and their co-workers (see Jacob and Monod, 1961, 1963; Monod and Jacob, 1961; Monod *et al.*, 1963; and also Brenner, 1965). An example of this type of system is again provided by *E. coli*,

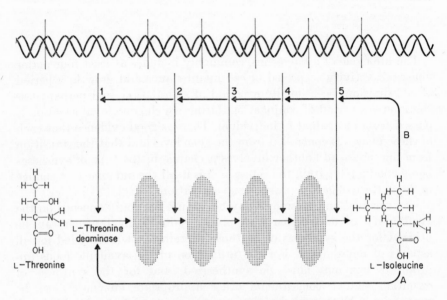

FIG. 1. The two feedback systems that control the synthesis of the amino-acid L-isoleucine in *E. coli*. The end product, when present in sufficient concentration, inhibits the activity of the first enzyme in the chain (**A**) and represses the further synthesis of all the enzymes (**B**). (Modified from Changeux, 1965.)

in which an excess of L-isoleucine not only inactivates the existing molecules of L-threonine deaminase but also inhibits the gene that specifies this enzyme so that no further synthesis occurs (Fig. 1).

The manner in which genes are activated or inactivated in response to conditions prevailing within the cell has been the subject of much recent research which has been integrated into what is known as the Jacob-Monod theory.

THE JACOB-MONOD THEORY

Briefly, Jacob and Monod have shown that the structure of an enzyme is specified by one gene, called a structural gene, and they have postulated that the rate of synthesis of the enzyme is determined by another gene, called a regulator gene. The primary product of a structural gene is a short-lived messenger RNA, which moves to the ribosomes to initiate the production of one, or at most a few, molecules of a particular enzyme. Since each RNA molecule is short-lived, a constant stream of such molecules must be produced if a significant quantity of the enzyme is to be synthesized and an adequate concentration maintained. A structural gene may act in isolation in this way, or it may form part of a group of two or several structural genes which act together as a coordinated unit to give rise to a number of metabolically sequential enzymes. Such a group, whether of one or several related genes, is called an operon, and it is now evident that the genes of one operon tend to lie adjacent to each other on the chromosome (see Demerec and Hartman, 1959).

The action of a regulator gene is primarily negative in that it represses the activity of a particular operon. Such a gene is regarded as being continuously active in producing a stream of mRNA molecules, which, it has been suggested, initiate the production of a stream of protein repressor molecules. Jacob and Monod (1961) have concluded that these repressor molecules operate "at a level where the information derived from several adjacent structural genes is still contained in a single, continuous, functionally integrated structure", and that the repression is in fact registered only in a short terminal region of the operon, a region that has been called the operator. This operator is regarded either as a gene or as part of the first structural gene of the operon sequence, and when an operator is repressed the whole attached operon ceases to function. More recently Jacob et al., (1964) have postulated the existence of another element, the promotor, which lies between the operator and the first gene of the operon.

Evidently the informational content of an operon must be read sequentially by a process that can only start either in the operator or in the adjacent promotor. De-repression of the operator evidently renders the DNA available for mRNA synthesis, a process that requires the catalytic assistance of the ubiquitous enzyme RNA polymerase (Wood and Berg, 1962). It is suggested that this enzyme attaches to a short length of the DNA molecule where the DNA-DNA base pairing has been temporarily broken, and so catalyses the production (probably on only one DNA strand) of a complementary mRNA molecule which transiently forms a DNA-RNA complex. Thus the synthesis of mRNA probably involves the unwinding of the DNA helix and the separation of the DNA strands in the operator, or promotor, region; the binding of RNA polymerase to one of the DNA strands; the stepwise advance of the RNA polymerase as the unwound and separated DNA zone advances along the length of the operon; and the stepwise synthesis and detachment of the mRNA chain.

Since this process must begin in the operator-promotor regions, the repression of these regions is all that is needed to inactivate the whole operon. On the present theory it is assumed that the DNA is inactivated when it is bound to some substance which prevents local uncoiling or local RNA polymerase action or both. It is possible that this substance is the repressor produced by the regulator gene, but it is equally possible that this repressor may merely be part of a more or less complex network of reactions which provides the connection between the repressor and the DNA. In either case this system may be simpler in the microorganisms than it is in the cells of higher organisms, which besides being diploid possess histones as an integral part of their chromosomes; in such cells it is commonly believed that histone binding may play some role in gene repression (but see p. 29.)

THE *lac* OPERON OF *E. coli*

In such bacteria as *Escherichia coli* each regulator gene continues to produce its repressor, which continues to prevent protein synthesis until that repressor is neutralized by the presence of some substance that has been called an effector or an inducer (Fig. 2). Such an effector

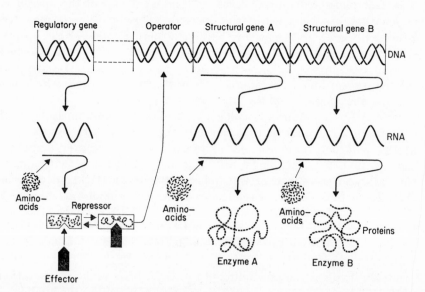

FIG. 2. The control of enzyme synthesis by gene activation and repression. The binding of the effector to the repressor allows it either to activate or inactivate the operator according to the type of system. (Modified from Changeux, 1965.)

may be a relatively simple food substance, and the classical example on which the Jacob-Monod theory is largely based has been provided by an analysis of the reactions of normal and mutant forms of *E. coli* in the presence of lactose. Populations were isolated of bacteria which can utilize lactose (which is not a normal food material) and of mutants which cannot. The mutants were of a number of types: those which synthesized structurally abnormal (and therefore inactive) forms of β-galactosidase, which is the enzyme needed to hydrolyse lactose to glucose and galactose; those which were unable to synthesize β-galactoside permease, which is needed in the cell wall to enable lactose

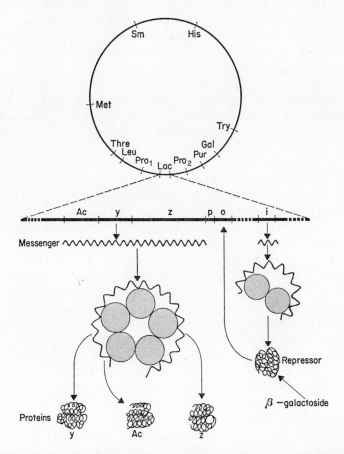

FIG. 3. The circular chromosome of *E. coli* showing the lactose region. This region is enlarged below and contains: *i*, regulatory gene; *o*, operator; *p*, promotor; *z*, gene for β-galactosidase; *y*, gene for β-galactoside permease; *Ac*, gene for β-galactoside trans-acetylase. The presence of β-galactoside inactivates the repressor and so activates the operon. (Reproduced with permission from Jacob, 1966.)

to be absorbed; those with an abnormal operator which prevented the whole operon from becoming functional; and those which, having an abnormal regulator gene, were unable to synthesize the normal repressor with the result that the operon continually synthesized mRNA at the maximum rate. In addition populations were isolated in which gene translocation had occurred, and it was these which showed that the operator must be adjacent to the first structural gene of the operon but that the regulator gene can function at a distance.

The details of the evidence are irrelevant here (see Jacob and Wollman, 1961; Hayes, 1964). The important conclusions are: that the structural genes specifying the enzymes β-galactosidase and β-galactoside permease, as well as possibly another enzyme, β-galactoside transacetylase (the function of which remains doubtful), form a natural group, the *lac* operon; that at one end of the *lac* operon is a short segment, the operator; that at a short distance and unconnected with the operator is the regulator gene, which specifies the repressor of the operator (in other known systems one regulator gene can control more than one operon); and that in the presence of the appropriate effector, which here is lactose, the repressor is inactivated and the operon is activated (Fig. 3).

CO-REPRESSION

The enzymes specified by the *lac* operon of *E. coli* are inducible enzymes, which means that they are not produced in any significant amount except in the presence of lactose, when the active repressor is inactivated by the effector.

However, a converse situation is also known in which the repressor is inactive except in the presence of the effector. An effector which activates a repressor has been called a "co-repressor", and in bacteria co-repression is as common a phenomenon as is induction. A well-known example is the inhibition by tryptophane of the synthesis of tryptophane-synthetase in *E. coli* (see Jacob and Wollman, 1961; Hayes, 1964). Tryptophane is an essential amino-acid; when it is present in the food it does not need to be synthesized, but when it is absent the enzyme tryptophane-synthetase is produced to catalyse the reaction indoleglycerol phosphate + serine → tryptophane + triose phosphate. In the absence of tryptophane the inactivity of the repressor permits the active synthesis of the enzyme; in the presence of the tryptophane the repressor is activated and synthesis ceases. Since this particular enzyme is constructed of two polypeptide chains, each specified by a separate gene, the operon consists of two genes.

A similar but more complex example of co-repression concerns the synthesis of the essential amino-acid histidine in *Salmonella typhimurium* (Ames and Hartman, 1963). The eight structural genes that specify the enzymes needed for the synthesis of histidine from phosphoribose pyrophosphate and adenosine triphosphate are grouped together to form the *his* operon, one end of which contains the operator. The repressor is active only in the presence of histidine in the food.

By exploiting gene control mechanisms of these types, a bacterial cell can respond appropriately to a wide variety of complex changes in its environment. It is, however, possible to devise experiments which make nonsense of these mechanisms. Thus artificial compounds closely similar to lactose can activate the *lac* operon although they cannot be broken down by β-galactosidase, and conversely the substance 5-methyl tryptophane can stop the synthesis of tryptophane although it cannot be utilized by the cell, which then dies of tryptophane starvation.

CONTROL BY ALLOSTERIC PROTEINS

In all the various systems so far adequately analysed, the regulatory part of the genetic mechanism "involves a system of transmitters (regulator genes) and receivers (operators), of specific cytoplasmic signals in the form of repressor molecules, which have the double property of recognizing a particular metabolite and a particular operator" (Jacob and Monod, 1961). The problem that immediately arises concerns the method whereby the repressor recognizes both the metabolite and the operator.

The repressor is commonly assumed to be a protein (see Monod *et al.*, 1963; Gilbert and Müller-Hill, 1966), the main reason being that it is activated or inactivated in the presence of small effector molecules, such as lactose and tryptophane, and that it is well known that during such combinations proteins may suffer dramatic changes in their properties. Proteins that are capable of such reactions are called allosteric proteins, and if the repressors are allosteric then they must possess at least two non-overlapping receptor sites. It is postulated that one of these, the active site, binds directly or indirectly to the operator mechanism, while the other, the allosteric site, binds to the effector substance. It is further postulated that when the effector binds to the allosteric site there is a change in the tertiary structure of the repressor protein which alters the active site. Thus in the presence of lactose the active site is inactivated, binding with the operator becomes impossible, and the *lac* operon becomes active; in the presence of tryptophane the active site is activated, binding with the operator becomes possible, and the tryptophane synthetase operon becomes inactive.

As a simple example of an allosteric transition (which, however, has no connection with genetic mechanisms), Monod *et al.*, (1963) refer to the reaction of haemoglobin with oxygen. Although the transition so induced is not completely understood "it is certain . . . that the binding of oxygen to a haem induces within the molecule a redistribution of charge, expressed as a discharge of protons by an acidic group". The allosteric repressor proteins and their stereospecific bindings are thus considered to be merely highly specialized manifestations of what seem to be the general properties of most if not all proteins. In this connection it is important to note that the effectors binding on the allosteric sites of the repressors may be quite unrelated to the substances, or to the products of the substances, which bind to the active sites.

MODIFICATIONS OF THE JACOB–MONOD THEORY

It must be emphasized that much of the operon theory outlined above is still unproved and in particular that the allosteric repressor control system is so far entirely hypothetical. Recently evidence has been accumulating of possible important deviations from the Jacob–Monod theory, and these have been briefly reviewed by Brenner (1965). Thus it has been found that although the close linkage of related genes into compact operons does often occur, as in the *lac* operon, dispersed operons may be equally common. An example of such a dispersed operon in *E. coli* is provided by the genes which specify the eight enzymes needed for arginine synthesis and which are scattered in five different locations on the chromosome. All the genes are repressed in the presence of arginine and it appears that each of the five groups possesses its own operator. It has even been shown that in some operons of closely linked genes, each gene may possess its own operator (Ramakrishnan and Adelberg, 1965).

There is also evidence in the case of the histidine-synthesizing enzymes that the actual co-repressor is not histidine itself, but that it is some substance whose production is evoked by the presence of histidine (Schlesinger and Magasanik, 1964). The mechanism controlling alkaline-phosphatase synthesis in *E. coli* is even more complex. Garen and Echols (1962) have suggested that the phosphate is not itself the co-repressor, but that it is converted by the product of one repressor gene (R_1) into an inducer which is then converted by the product of a second repressor gene (R_2) into a co-repressor.

These are, however, merely modifications of the basic control theme and many such modifications may be expected to occur, especially in the cells of higher organisms. The one serious contradictory point raised by Brenner (1965) concerns the very existence of regulator genes and of allosteric repressor proteins, and he reviews possible alternative hypothetical control mechanisms which are suggested especially by the work of Stent (1964) and of Hartwell and Magasanik (1963). At this moment it is impossible to weigh the value of these hypotheses one against the other, and although it remains unsubstantiated the Jacob–Monod repressor hypothesis is tentatively accepted here as the simplest and neatest explanation of the facts at present available. Important as the problem obviously is, for present purposes an understanding of the precise nature and degree of complexity of the connection between the inducer (or co-repressor) and the relevant operon is less important than the fact that such a connection obviously exists, and that by its exploitation a bacterium is able to change its pattern of gene activity from moment to moment to meet the changing environmental circumstances.

DIFFERENTIATION IN BACTERIA

The genetic mechanisms whereby a bacterial cell controls the rate of synthesis of its various enzymes has been discussed in some detail merely to serve as an introduction to the question of differentiation in these organisms. As mentioned above, the term differentiation was originally applied to the process of tissue formation in the higher organisms, but it seems most improbable that this process can represent anything more than an evolutionary development of some simpler type of cellular control mechanism.

It was Monod and Jacob (1961) who first emphasized the apparent similarity between the bacterial control mechanisms, which limit the expression of the genetic potentialities, and the control mechanisms of multicellular organisms, which during embryonic life also limit the expression of the genetic potentialities of the cells of the forming tissues. The one apparent difference between the two situations is that in bacteria gene expression readily changes when the inducers are added or withdrawn, while in higher organisms differentiation, once induced, is relatively stable. It is, however, reasonable to suggest that these two different situations may represent the two extremes of the possible range. Evolutionary selection in bacteria would be expected to favour a type of differentiation that permits easy reversibility, since it is obviously advantageous for them to be able to respond rapidly and appropriately to environmental change; in higher organisms evolutionary selection would be expected to favour the type of differentiation that produces the maximum stability in tissue function.

In attempting to define the essence of differentiation Jacob and Monod

(1963) have proposed "that two cells are differentiated with respect to each other if, while they harbor the same genome, the pattern of proteins which they synthesise is different". Viewed in this way the bacteria show a kaleidoscopic range of differentiation which is necessary to maintain their intracellular homeostasis, while the tissue cells of higher organisms, bathed in a medium the composition of which is itself homeostatically controlled, have little need of this ability.

PATHOLOGICAL DIFFERENTIATION

The fact that the control of both bacterial and tissue differentiation is vested ultimately in the DNA, has allowed the development of that peculiarly intimate form of parasitism shown by the DNA viruses. Viral DNA is able to subvert the cellular machinery for its own purposes and to direct syntheses which result in pathological, and sometimes fatal, forms of cell differentiation. Again one of the classical cases is provided by *Escherichia coli*, which is attacked for instance by a virus that has been designated T_4 (see Edgar and Epstein, 1965). In this case so great is the activity of the invading DNA that in only about 20 minutes some 200 virus particles have been formed and the bacterium is destroyed.

Still more interesting, however, is the type of virus which, when its DNA enters a bacterium, may either behave virulently in the manner of the T_4 virus or may instead join with the host chromosome and behave as though it had become an integral part of the host DNA. A virus sub-dued in this way is known as a provirus, and by directing the production in the bacterium of abnormal proteins it gives rise to a specific pathological form of differentiation. One aspect of this is commonly an unusual coating on the bacterial surface which can be detected by immunological methods, while another is the suppression of any further viruses of the same species which may gain entry. This suppression is apparently one expression of a specific proviral regulatory system which acts to prevent the explosive growth typical of the virulent viral state.

CYCLIC CELLULAR ACTIVITIES

In nature the patterns of bacterial gene activity are, no doubt, forever changing in step with the changing cell environment. The transience of these types of differentiation and the speed of transition from one type to another is due, in the first place, to the rapid response of the genes to their effector substances, and in the second place, to the instability of the mRNA, which, when the effector is removed, may disappear within two minutes. However, bacteria periodically undergo more ambitious forms of differentiation, which are also more stable in that once they have started they must pass to completion through a sequence of processes which may sometimes take more than two hours. These are the periodically occurring phenomena of cell division, which in ideal conditions may occur two or three times an hour; of sporulation, which takes place when conditions deteriorate; and of conjugation, which in origin may be pathological (see pp. 23 and 25). Being stable all-or-none reactions occupying relatively long periods of time, these three periodic phenomena are clearly more complex than the effector reactions so far considered, and they must depend on control systems of considerably greater intricacy. However, there is no evidence to suggest that these control systems are in any way fundamentally different from those of the simpler effector reactions, and it seems certain that all of them must depend on the periodic transcription of specific regions of the genome in response to the presence of specific chemical messages.

CELL DIVISION

It is clear that in any normal bacterial cell the process of transcription of DNA from DNA is as strictly regulated as is that of the transcription of mRNA from DNA, and it is also clear that some specific regulatory mechanism must exist to determine both the time of onset and the course of chromosome duplication. One clear indication of the existence of such a mechanism comes from studies of proviruses. Although in a provirus the repressed viral DNA does not specify the enzymes needed to initiate and control DNA duplication, it does respond together with the host DNA whenever such enzymes are produced within the bacterium. Consequently whenever the infected bacterium undergoes cell division, so also does the provirus. It is many years since it was first shown that extracts of *E. coli* contain an enzyme called DNA polymerase which is necessary for DNA replication (Lehman *et al.*, 1958), and it is now evident that other enzymes are involved in the reaction as well (Richardson *et al.*, 1963). DNA polymerase is evidently a generalized enzyme, as is shown by the activity of extracts of *E. coli* in promoting replication, for instance, in the DNA of *Bacillus subtilis*.

In *E. coli* it has been established both by genetic evidence (Jacob and Wollman, 1961) and visually by autoradiography (Cairns, 1963) that there is only a single chromosome which forms a closed ring. This chromosome has the form of a spiral formed of two complementary DNA strands, and when this duplex structure is fully extended it has a length of about 1·4 mm. Since it is not complexed with protein it is extremely fragile. DNA replication continues throughout almost the entire cell cycle (Maaløe, 1961), and the indications are that this process always begins at a particular point on the ring. The section of DNA at this point has been called the replicator (see Jacob *et al.*, 1963), and from it the process of DNA synthesis always travels in the same direction. It has also been shown that the synthesis of some particular substance or substances is necessary for replication to begin but not for it to continue (Maaløe, 1961, 1963). It is obviously tempting to suggest that this substance may control DNA synthesis in the replicator segment of the chromosome in much the same way as an effector is

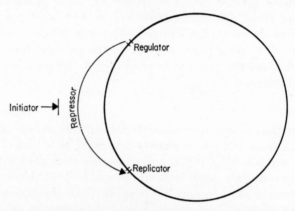

Fɪɢ. 4. The regulation of DNA duplication in the circular chromosome of *E. coli*. It is suggested that a regulator gene specifies a repressor, which inactivates the replicator region (or operator), unless it is itself inactivated by an effector, or initiator.

believed to control its appropriate operator. The only difference between the two situations is that chromosome duplication involves the complementary replication of both DNA strands by the incorporation of deoxyribonucleotides, while mRNA synthesis seems to involve the complementary replication of only one DNA strand by the incorporation of ribonucleotides. On this view the replicator can be regarded as an operator while its dependent operon for DNA synthesis is the entire chromosome. It can be suggested that, as usual, a specific repressor is probably synthesized under the direction of a regulator gene (see Fig. 4; and Jacob *et al.*, 1963), that this repressor may normally be active, and that, to become inactivated, it must react with some unknown trigger, substance, or effector, coming perhaps from the cell membrane. On this view the cyclic initiation of cell division would be determined by the cyclic production of the trigger effector, which would itself be formed in response to conditions both within and without the cell.

All this is mainly hypothesis, but some supporting evidence has been assembled by Jacob *et al.* (1963), especially from studies of mutations in provirus DNA. Situations have been found in which the provirus chromosome lacks a functional replicator, and in this situation DNA replication becomes impossible unless, as sometimes happens, the proviral DNA becomes joined to the bacterial DNA. In such a combination the proviral DNA replicates synchronously with the bacterial DNA, the whole process evidently being initiated by the bacterial replicator. Obviously mutations leading to this type of abnormality cannot be studied in the bacterial chromosome since, when they occur, they must lead rapidly to the death of the cell.

Although relatively little is yet known of the details of chromosome duplication in bacteria, enough is known to indicate clearly that some control mechanism does exist, that it is mediated through changes in gene activity, and that it must be initiated by the production of some substance which acts as a trigger. Using the definition given above, it is also clear that a bacterium involved in DNA replication and cell division can be properly regarded as being in a particular state of differentiation. It must be emphasized, however, that this replicative state of differentiation does not interfere with, but rather overlaps, the varying states of vegetative differentiation that occur in response to environmental changes. Cell growth continues as DNA replication proceeds, and indeed Ryter and Jacob (1963) have obtained evidence suggesting that the physical separation of the daughter chromosomes may be the result of the growth of the cell surface. They believe that the chromosome ring is attached by the replicator to the cell surface, that after chromosome duplication begins there are two points of attachment, and that as these become separated by cell elongation a new cell wall is laid down between them. Certainly there appears to be no trace in bacteria of any spindle apparatus associated with chromosome separation.

The high rate of chromosome duplication in bacteria is evidently an adaptive feature which enables these organisms to survive their extremely high rate of death. To illustrate this it has been pointed out that the progeny of a single bacterium dividing every twenty minutes in ideal conditions could in two days achieve a mass greater than that of the world (Srb *et al.*, 1965).

SPORULATION

The various phases of biochemical and structural differentiation found in growing and dividing bacterial cells continue unimpeded only so long as a rich supply of nutrients is available. Whenever the supply of nutrients is limited the cells embark on a series of changes leading to the formation of structures which have been variously classified as endospores, cysts, and conidia, but which here will simply be called spores. This phenomenon has been reviewed in detail by Halvorson (1965), who has emphasized that it must be regarded as a form of differentiation.

Spore formation in the genus *Bacillus* involves the division of the cell into two compartments, a "fore-spore" which matures into a dormant and resistant spore, and a "sporangium" which evidently continues to behave like a vegetative cell. It is only in the maturing "fore-spore" that the syntheses typical of sporulation occur. This process involves both alterations in the already existing metabolic pattern and the production of new types of macromolecules. In the genus *Bacillus* major metabolic changes have been described in the methods of glucose utilization and of electron transport. However, it is the synthesis of new types of macromolecules that are more significant in the present context. Such syntheses lead, for instance, to changes in the constitution of the cell wall, which can be demonstrated by the detection of novel antigenically active substances and by the obvious fact that the wall becomes much more impervious. They also lead to changes in a number of enzymes, such as a catalase, from heat-sensitive to heat-resistant forms, and there is even an indication that some of the new enzymes may be more radio-resistant (Rowley and Newcomb, 1964). These are all syntheses which clearly are appropriate preparations for a phase of life in which resistance to environmental extremes is essential. The degree of resistance achieved is illustrated by the ability of *Bacillus* spores to survive for some two or three centuries (Sneath, 1962).

Evidence that these preparations for spore formation are gene-directed has been assembled by Halvorson (1965). The new enzymes that appear during sporulation are synthesized in response to the appearance of new types of mRNA that are specific to sporulation, and direct evidence of gene control has been obtained, for instance, with *B. subtilis* (Schaeffer *et al.*, 1964). In this organism at least six successive morphological stages have been recognized in the course of sporulation, and mutants have been isolated in which sporulation is able to proceed only as far as one or other of these stages. It is thus clear that the morphological, and therefore the biochemical, changes leading to the mature spore are under a series of separate gene controls (Fig. 5). Other similar examples are quoted by Halvorson (1965), who concludes that "sporulation can thus be viewed as a process in which, following the inactivation of the vegetative genome, a large number of spore specific components are synthesized sequentially", and that "the spore genome contains a large number of genes which are probably distributed in a number of operons that are transcribed in a specific order". Halvorson speculates that the sporulation genes of *B. subtilis* must number at least 100, and many mutations both of structural and of regulatory genes have already been recognized.

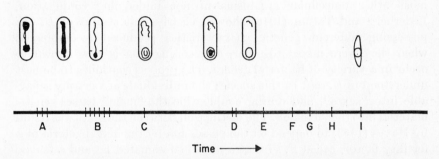

FIG. 5. The sequential morphological and biochemical changes during sporulation in *Bacillus* (time scale A-I about 7 hr). *A*, commitment to sporulation with production of sporulation factor; *B*, acetate oxidation; *C*, ribosidase; *D*, S-S enzymes; *E*, presence of dipicolinic acid and uptake of calcium; *F*, heat resistance; *G*, alanine racemase; *H*, lytic system; *I*, release of spores. (Reproduced with permission from Halvorson, 1965.)

This process of sporulation resembles that of DNA replication and cell division in that, once begun, it always proceeds to completion. It may, therefore, be suspected that sporulation must begin in response to some trigger substance, which is produced in adverse conditions, and which, as in the case of the activation of DNA replication, may inactivate some otherwise active repressor that is constantly present in the vegetative cell. Recently Srinivasan and Halvorson (1963) have isolated a sporulation factor, of low molecular weight, which seems to possess the necessary characteristics of a trigger substance. When it is added to bacteria in the exponential phase of vegetative growth, sporulation is induced.

When a spore is fully formed all the mRNA is rapidly lost, and during the resting period few, if any, such molecules are produced. When ultimately the spore finds itself in the presence of abundant nutrients, the vegetative part of the genome again becomes active, apparently in response to the production of another trigger substance. Within only a few minutes mRNA and enzyme synthesis begins, and the proteins typical of the vegetative state appear in an ordered sequence. Meanwhile the spore elongates and after about 100 minutes DNA replication recommences. It thus appears that both sporulation and germination, although opposite processes, are basically similar, in that they each depend on the activation of previously quiescent genes by some trigger substance and the transcription of these genes in an ordered sequence.

CONJUGATION

Although for a long time it was believed that bacteria differed from most other unicellular organisms in not undergoing conjugation, Lederberg and Tatum (1946) showed clearly that strains of *E. coli* possessing different genetic characteristics produced recombinants when they were mixed together, and conjugation is now known to occur in a variety of bacterial genera. The process continues to be best understood in *E. coli*. In this species the individuals are sharply separable into "sexes", and during conjugation the "male" passes genetic material into the "female". The distinction between the sexes was found by Hayes (1953) to depend on the possession by the male bacteria of a fertility factor, called F, so that a male is designated F^+ and a female F^-. This F factor has been identified as a short segment of DNA which commonly lies free in the cytoplasm but which may combine with the circular chromosome. When free in the cytoplasm the F factor replicates autonomously, but when linked with the chromosome it replicates with it.

Conjugation is evidently relatively rare in wild-type populations of *E. coli*, occurring perhaps in only about one in a million random collisions. As seen through the electron microscope, the process involves the junction of the two bacteria by a slender conjugation tube through which genetic material passes. When the F factors are free in the cytoplasm then one of these is usually the only genetic material that passes into the female cell, which is then changed from the F^- to the F^+ form. Such differentiation of a female into a male cell is a relatively rapid process, being completed in about thirty minutes (Valentine, 1966). Only rarely in this type of conjugation does any part of the main chromosome also penetrate into the female.

When, however, the F factor is linked to the chromosome of the male cell, the course of conjugation is different and the probability that random contact between cells will lead to conjugation is increased about a hundred fold. Males of this type are therefore called supermales and form what is known as the high frequency combination strain, or the Hfr strain. Individuals of this type often appear in F^+ populations, apparently as the result of the spontaneous linking of the F factor with the chromosome. Jacob and Brenner (1963) have suggested that when a supermale conjugates with an F^- individual, DNA duplication is initiated in it. When the conjugation tube has been formed one of the circular chromosomes breaks at the point of insertion of the F factor, and the chromosome thread then begins to pass into the F⁻ individual. Evidently this moving thread is easily broken, and when this happens conjugation comes to an end. It follows that commonly only parts of the male chromosome enter the female cell. However, it sometimes happens that the entire male chromosome manages to pass through the conjugation tube, and in this case the whole process takes about two hours. The last segment of genetic material to enter the female cell is the F factor, which then converts the cell into a male.

There are no regular haploid and diploid phases in bacteria, but the "zygote" can perhaps be regarded as "diploid" in respect of whatever fragment of the male chromosome it has received. It then commonly happens that, by recombination, genes received from the male are substituted for similar genes in the female chromosome. This produces a chromosome with a completely new gene pattern, and it is this mixing and recombining of the genetic potentialities of the population that may perhaps confer some slight advantage. However, since bacteria are essentially haploid there is no question of the spread of hidden recessive genes through the population, which is one main consequence of conjugation or fertilization in diploid organisms.

The DNA of the F factor has sometimes been considered to be a supernumary chromosome and sometimes a special form of virus, and because it induces maleness even when separated from the main chromosome it has also been called an episome and cited as a form of extra-chromosomal inheritance (see Jinks, 1964). Its similarity to a virus is striking (Hayes, 1966). Like the F factor, the genetic material of a temperate bacteriophage when free in the cytoplasm replicates rapidly and antonomously, but when linked with the chromosome it behaves like a provirus and replicates only once per division cycle. Also like a provirus the integrated F factor is able to exclude from the cell any free replicating forms of the F factor that attempt to gain entry (Jacob and Wollman, 1961).

Whatever the actual status of the F factor may be, however, it is clearly composed of those genes that control sexual differentiation and therefore conjugation. As a result of the presence of these genes in either the F⁺ or the Hfr strains, substances are synthesized, first, to alter the properties of the cell surface so as to permit adhesion to F⁻ individuals (the altered nature of the cell surface is also shown by changes in antigenicity and by increased susceptibility to penetration by certain bacteriophages), and second, to allow the development of the necessary mechanism to drive the opened chromosome through the conjugation tube. The induced changes in the cell surface are evidently present at all times, so that this aspect of male differentiation is permanent. It would be interesting to discover whether the enzymes that control the actual process of conjugation are also present at all times or whether their synthesis is induced in response to some trigger mechanism which is itself activated when the surfaces of F^+ (or Hfr) and F^- cells make contact.

Conclusions

It is generally agreed that the bacteria possess the simplest and probably the most primitive type of cell structure surviving today, and they are by far the most ancient group of which fossil record remains. It appears that the main armament used by these simple organisms to combat the constantly varying and often hostile environment is their ready ability to adopt a physiological state appropriate to the challenge of the moment. The kaleidoscopic range of physiological states of which they are capable is controlled by the varying concentrations of a range of effector substances in the cytoplasm. In the simplest situations the concentration of any effector substance inside the cell is dependent on its concentration outside the cell, but in more complex situations specific effector substances may be synthesized inside the cell. The evidence for these conclusions, obtained by the application to bacteria of classical genetical methods, has been assembled to support what is known as the Jacob–Monod theory of gene control (Fig. 2).

THE ESSENCE OF DIFFERENTIATION

One outcome of these genetic studies of bacteria has been the realization that when two organisms possessing the same genome show different effector-controlled patterns of gene activity, and therefore of enzyme synthesis, each must be regarded as differentiated in respect of the other. This extension of the meaning of the term differentiation is deliberately intended to imply that the unstable and often transient states of differentiation found in bacteria are directly comparable to the stable states of differentiation found, for instance, in the tissues of the higher animals. The essential similarity lies in the fact that both depend on a reaction whereby certain groups of genes are activated or repressed in response to the presence of specific effector substances. It appears that the highly labile states of differentiation found in bacteria and the highly stable states of differentiation found, for instance, in the tissues of a mammal may merely represent the two extremes of the range of differentiation, and that many intermediate partly stable types exist. It is clear that lability is of outstanding importance to micro-organisms while stability is essential in mammalian tissues.

SYSTEMS OF DIFFERENTIATION

In bacteria the simplest form of the Jacob–Monod model, in which the presence or absence of a single effector determines the synthesis of a single enzyme, can be regarded as a physiological unit of differentiation. It represents a single regulatory circuit whereby the effector reacts with the product of a regulator gene to determine the activity of a single operon.

However, it is obvious, at least in theory, that any number of such units, or regulatory circuits, can be interlinked to form systems of differentiation with any degree of complexity and with any desired properties. The most complex systems of differentiation so far described in bacteria are those that lie behind cell division, sporulation, and conjugation. Each of these processes evidently depends, first, on the activation of a trigger mechanism, and second, on a chain reaction which involves the activation in correct sequence of a number of interrelated operons and the consequent synthesis of an even larger number of special enzymes.

The only limit to such elaboration is set by the number of genes that are available in the genome, and in bacteria this number is relatively small. This may be the main reason why the bacteria have remained for so long a sterile evolutionary backwater and why the highest degree of complexity achieved by this cell type is represented only by the filamentous blue-green algae.

THE ORIGIN OF SEX

Conjugation is a particularly interesting process in which the essential role is played by the F factor, or sex factor, which is composed only of DNA and which confers the quality of maleness. There are good reasons for regarding this sex factor as a temperate bacteriophage with a novel mode of infection, and as a result of conjugation "the character of maleness spreads through the female population like an epidemic" (Hayes, 1966). The "disease" of maleness can also be "cured" by treatment with acridine orange (Hayes, 1967). The question of the value of this association, both for the bacterium and for the virus, obviously arises. For the bacterium there seems relatively little to be gained by any exchange of DNA and recombination of genes. Being haploid, bacteria can retain no recessive mutations for later expression in changed circumstances, as occurs in diploid organisms. Hayes (1966) has concluded that the conjugation methods "in bacteria are rudi-

B

mentary and appear to operate very inefficiently under natural conditions", and that it "seems unlikely that sexuality *per se* can have played a major role in bacterial evolution".

However, for the viral sex factor itself the advantages are obvious since, by avoiding killing its host and by inducing conjugation, it has developed a particularly elegant method of ensuring a safe passage from one host cell to the next. Thus sex and conjugation in the bacteria may be regarded as expressions of a pathological state of differentiation induced in the male cell by the presence of parasitic viral DNA.

It must be emphasized that this viral DNA also induces the "disease" of maleness in bacteria other than *E. coli*; it readily infects a variety of genera and species of intestinal bacteria, which then also show conjution. However, as far as is known, only one other genus, *Salmonella*, possesses a chromosome with "sufficient genetic similarity to the sex factor to permit its insertion" so as "to generate Hfr males and promote chromosomal transfer" (Hayes, 1967). It must also be added that this particular sex factor is not the only viral DNA that is able to induce conjugation. This method of viral infection seems to have been evolved on many separate occasions.

Although conjugation may be relatively unimportant to the bacteria themselves, it is possible that their close, and perhaps ultimately permanent, relation with some primitive viral sex factor may have provided the basis from which all other forms of sexuality have evolved. If this is true this may be the most important single evolutionary contribution that has come from the bacteria, and indeed without the development of the habit of conjugation it is difficult to see how any advanced forms of multicellular organisms, most of which rely primarily on sexual reproduction, could have arisen (see also p. 52).

THE BACTERIA AND EVOLUTION

The bacteria may have been limited in their evolutionary potentialities by the small size of their gene complement, and have given rise in the early pre-Cambrian only to the blue-green algae on the one hand and to their own obligate parasites, the viral bacteriophages, on the other. However, it is obvious that in structure and function, their cell is closely similar to the more advanced type of cell possessed by all plants and animals. Ancient and primitive as they are, the bacteria already possess all the basic cellular attributes, including the typical methods of DNA coding, of RNA transcription and enzyme synthesis, of gene control leading to differentiation, and of conjugation leading to a form of sexuality. It is therefore possible that the early bacteria, or some closely related organisms, also gave rise to the higher type of plant and animal cell. There is indeed good reason to believe that all the complexities found in the cells of higher organisms are merely elaborations of the basic principles of cell organization and control that were already established in these simple cells.

CHAPTER 3

Unicellular Plants and Animals

It has been suggested that the most ancient surviving cell type is that of the bacteria, which gave rise on the one hand to the viruses and on the other to the blue-green algae. This cell type has been called prokaryote to distinguish it from the more highly organized eukaryote cell, which is typical of all other organisms from the unicellular algae and protozoans to the higher plants and animals. It is possible that the eukaryote cell evolved from the prokaryote cell, but even if this did not happen it is still evident that they are so closely related that they must have been derived from a common ancestor.

The eukaryote cell is characterized by the possession of a nuclear membrane and therefore of a sharp separation of nucleus from cytoplasm; by a nucleus containing at least two chromosomes (diploid) in which the double DNA thread is complexed with basic protein (usually histone; see Iwai, 1964, and Prescott, 1964) and to which are attached one or more nucleoli (see Grimstone, 1961); and by the presence in the cytoplasm of a range of organelles, which are regions where particular metabolic activities have become centralized and which are consequently visible microscopically. These organelles include, for instance, a form of endoplasmic reticulum with ribosomes (see Scherbaum and Loeffer, 1964), chloroplasts (in plants) and mitochondria (see Sager, 1964; Goodwin, 1964) which together contain the main energy-transforming systems, and the centriole which is involved in mitosis and which is seen only in animal cells, although its presence is inferred in plant cells (Mazia, 1965).

One of the most outstanding characteristics of the eukaryote cell is its habit of division by mitosis. This complex process begins after DNA duplication is complete and involves, in prophase, the contraction or condensation of the chromosomes, in metaphase, the final synthesis and arrangement of the spindle apparatus which links the chromosomes to the two centrioles and these to the two poles of the cell, and in anaphase, the movement of the chromosomes, by some unknown means, towards the centrioles. It is possible that all these processes are in some degree dependent on the presence in the chromosomes of histones, which may provide points of attachment to the spindle whereby the chromosomes are assembled and separated and which may be subject to those changes that cause the prophase condensation.

It is certainly probable that one of the main functions of the chromosomal histones is to provide a framework on which the DNA thread can be folded and stabilized in some carefully ordered manner (see Mazia, 1965). The DNA thread is much longer than even the extended interphase chromosome, and its ordered folding becomes essential in any cell in which it exceeds a certain critical length. Beyond this length a free DNA thread would be impossibly unwieldy and would be in danger of becoming entangled, especially when more than one chromosome was present in the cell. It is known that while the naked DNA of the single bacterial chromosome, containing perhaps a thousand genes, may when fully extended have a length of about 1 mm, the DNA complexed with histone in a single animal chromosome, containing more than ten thousand genes, may have an extended length exceeding 1 cm. A single human cell with forty-six chromosomes may contain more than a million genes and the total length of the extended DNA may exceed one metre (Nirenberg, 1963). It is certainly reasonable to suggest that chromosomal DNA must be stabilized between divisions and organized during division in any cell that contains and handles more than a relatively small amount of DNA, and that it was the evolution of the eukaryote type of chromosome that allowed the vast increase in the informational content of the cell on which all evolution above the simplest unicellular level has depended. The reasons for the complete dominance of this cell type over the prokaryote are therefore obvious.

Since the unicellular eukaryotes first evolved, their descendants have tended to follow four main evolutionary trends: first, the conservative trend followed by the ancestors of many present-day unicellular algae and protozoans, and especially those that are included among the flagellates and rhizopods (however, some unicellular algae have reached relatively enormous sizes); second, the trend towards simpler cell structure (and often smaller size) in parasitic forms, the most specialized of which are sporozoans; third, the trend towards more complex cell structure (and often larger size) followed especially by the ciliates; and fourth, the trend towards the multicellular condition which has been followed on many occasions and which has led to a wide variety of modern organisms ranging from the cellular slime moulds to the mammals.

UNICELLULAR ALGAE AND THE SIMPLER PROTOZOANS

Relatively few unicellular organisms have been intensively studied from the point of view of differentiation, but these few range from the unicellular algae to the social amoebae (see p. 45). It is already obvious that all unicellular algae and protozoans from time to time pass through certain well-defined phases of differentiation.

Unlike the situation in the bacteria, little is known of changes in the pattern of enzyme synthesis due to changes in the nature of the food supply, but it is certainly clear that the genes and enzymes concerned with growth in the vegetative phase must periodically give place to genes and enzymes that are concerned with the sequence of changes that underlie DNA duplication and mitosis.

MITOTIC ACTIVITY

It is widely believed that when a cell begins preparations for cell division there is an abatement of vegetative activity, and this suggests that synthesis for mitosis tends to be an alternative to synthesis for cell growth. Thus, in the case of the unicellular alga *Chlorella*, Tamiya (1963) has spoken of the "competitive relation between protein synthesis and the formation of the division-inducing substances". This is in contrast to the situation in bacteria, where preparations for cell division accompany cell growth and where there is no mitotic spindle to be synthesized. It may be concluded that in all unicellular algae and protozoans in which synthesis for mitosis is a major task, there must tend to be a reciprocating activation and repression of the genes for vegetative growth and the genes for cell division.

It is interesting to find that when unicellular organisms change from vegetative syntheses to preparation for mitosis this change is sometimes triggered by an environmental variant. Thus Scherbaum and Loeffer (1964) have reviewed evidence that in fifteen green species of the flagellate *Euglena* cell growth occurs during the hours of daylight and mitosis is confined to the hours of darkness; in five colourless species this rhythm is absent and mitosis occurs at irregular intervals. The colourless flagellate *Astasia* can be trained to react to temperature oscillations with growth occurring during 16 hr at 15° C and mitosis during 8 hr at 25° C, and similar results have been obtained with *Amoeba* in which most mitoses also occur during the warmer period.

It seems clear that these environmental changes do not directly activate the mitosis genes and repress the vegetative genes. Rather it is probable that they set in motion some possibly complex mechanism which determines some critical change in the cytoplasm, possibly through the production or unmasking of a single effector substance. That preparations for mitosis begin in response to some cytoplasmic change is supported by the fact that in all multinucleate cells the nuclei always enter mitosis in unison. This is seen, for instance, in the rhizopod *Pelomyxa*, which may have up to a thousand nuclei (Kudo, 1947, 1951).

ENCYSTMENT

Another form of differentiation commonly found from time to time in unicellular organisms is the resting phase of encystment, which is essentially similar to sporulation in bacteria. It is commonly believed that it is adverse conditions which induce the formation of the protective cyst, and the habit no doubt enables the successful survival and widespread distribution of a large variety of species. In a number of algae and protozoans the actual trigger to encystment has been determined as low temperature, evaporation, change in pH, low oxygen content, or accumulation of metabolic products (Kudo, 1966), and to such environmental changes the cell reacts by changing its enzymic activities so as to produce a cyst wall that in green algae is usually cellulose and in colourless protozoans is usually chitin. The encysted cell usually loses such features as flagella and sometimes it may remain quiescent for years. Ultimately some further environmental change, such as the addition of water or oxygen, or even the presence of particular bacterial species (Crump, 1950), causes the organism to break out from the cyst, often by the synthesis of a specific enzyme, and to resume its vegetative activities.

Encystment of this type is particularly common among parasitic sporozoans, and in them the breaking of the cyst usually only takes place after entry into a new host. In many sporozoans, such as *Monocystis* and *Eimeria*, encystment only occurs after conjugation so that it is the zygote that enters the resting phase.

CONJUGATION

Yet another variation in gene activity is that which evidently precedes conjugation, and this method of exchanging genetic information is widespread among unicellular algae and protozoans. The special gene-directed syntheses that must occur before any organism can undergo conjugation are illustrated by the flagellate *Chlamydomonas*. Sager (1964) has described how, in *C. reinhardi*, normal individuals, which are all haploid, prepare for conjugation whenever they are starved for nitrogen. When this happens they lose their ability to initiate syntheses for mitosis, which is itself clear evidence of a change in the pattern of gene activity. In this species there are two mating types, which can be signified as + and −. They look alike but they are separable by their behaviour since only a + cell will mate with a − cell. These mating types are determined by a single pair of genes, and being haploid each individual possesses either one or other of these genes. Also, being haploid, no reduction division is necessary and each individual differentiates directly into either a + gamete or a − gamete. It now appears that one or both of these gametes produce substances that attract the other (see Wiese and Jones, 1963). The next step is the production of enzymes that digest the cellulose cell walls to allow the fusion of cytoplasm and nuclei and the production of a diploid zygote, which then enters the two reduction divisions to produce four new haploid individuals.

This long and complex phase of activity evidently involves the activation, in appropriate sequence, of a series of conjugation genes that normally are repressed, and the simultaneous repression of other groups of genes that would otherwise be active. The elaboration of such a pattern of conjugation genes was evidently basic to the evolution of a mechanism of gene transfer, which is considerably more complex than that found in the bacteria and which involves a sequence of cell activity from the diploid condition through meiosis to the haploid gamete. It is probable that more than one trigger mechanism must exist within this sequence.

POLYMORPHISM

Although the common phases of differentiation shown by the unicellular algae and the protozoans are those described above, many genera are also polymorphic. Well-known examples include dimorphic foraminiferans such as *Polystomella*, in which a multinucleate microspheric form alternates with a uninucleate macrospheric form; in *Discorbis opercularis* the microspheric form also has a shell with a clockwise rotation of the chambers, while the macrospheric form is anti-clockwise (see Kudo, 1966). These two forms are related to the conjugation cycle of these genera, the macrospheric form producing flagellate gametes which after fusion in pairs develop into the microspheric form.

Another example of polymorphism which is less obviously related to conjugation is shown by various genera of trypanosomes. In any one species the body form may vary widely, the differences being especially obvious between individuals present in the warm-blooded host and those present in the cold-blooded host (see Kudo, 1966). Some attempts have been made to analyse these differences *in vitro* (see Guttman and Wallace, 1964): in *Trypanosoma mega* the morphology of one form seems to be determined by a serum factor, probably a globin, while in *Trypanosoma conorrhini* and *Schizotrypanum cruzi* higher temperatures induce the form found in mammals and lower temperatures the form found in the invertebrate host. Whether these changes are the result of the direct action of the environment on the enzyme systems or whether they are mediated through altered gene action is unknown.

GENE CONTROL OF DIFFERENTIATION: *Acetabularia*

In almost all the situations described above the involvement of the genes can only be inferred; no direct evidence that the DNA→mRNA→ enzyme pathway is involved is available. However, strong support for the theory that changes at the gene level are directly responsible for the varying forms of differentiation in at least some unicellular algae comes from the extensive observations that have been made on *Acetabularia*, which is a giant unicellular marine alga. In suitable conditions it takes about three months to produce an upright stalk some 3–5 cm long, at the top of which are formed "whorls of deciduous laterals and finally a persistent cap" with a diameter of up to 1 cm (Fig. 6; and see Haemmerling, 1963). This complex structure is the fruiting body within which the haploid gametes are ultimately formed. However, during the growth of the stalk and of the fruiting body the single diploid nucleus remains in the basal rhizoid, which fixes the organism to the substratum, and it is this which makes *Acetabularia* so valuable for studies on the control of differentiation. The stalk can easily be cut so that the cell is divided into nucleate and anucleate parts, the morphogenetic potentialities of which can then be determined. In particular it can be discovered whether the formation of the fruiting body can

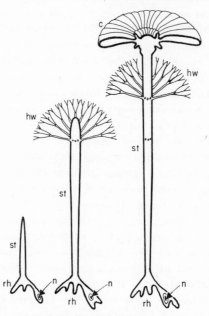

FIG. 6. The vegetative development of *Acetabularia*. Note the nucleus *n* remaining in the basal rhizoid *rh*; *st*, stalk; *hw*, hair whorl; *c*, corona. (Reproduced with permission from Brachet, 1967.)

continue in the anucleate part by the activation of already existing enzymes or whether it depends on *de novo* syntheses initiated by the nucleus.

The main results of this work are clear-cut and have been summarized by Haemmerling (1963) as follows: the continued growth and differentiation of the stalk and fruiting body are due to the action of "morphogenetic substances" which are continually produced by the nucleus; these "morphogenetic substances" are carriers of DNA-specified genetic information; and any temporary continuation of differentiation in an anucleate stalk depends on the quantity of these "morphogenetic substances" present in the cytoplasm at the time of separation from the nucleate rhizoid.

Supporting evidence comes from experiments in which the nucleus from one species of *Acetabularia* is inserted into the anucleate stalk of another species. The fruiting body then produced is always typical of the species from which the nucleus was taken. If two or more nuclei from different species are inserted, the resulting fruiting body is intermediate in form. Since all the nuclei were carefully washed clean of cytoplasm this indicates clearly that the details of morphogenesis are dictated by the nucleus and not by the cytoplasm.

The next critical question concerns the nature of the "morphogenetic substances". Stich and Plaut (1958) have shown that when nucleate and anucleate pieces of *Acetabularia* are treated with ribonuclease, all protein synthesis for morphogenesis ceases, and that when the ribonuclease is removed, protein synthesis recommences only in the nucleate pieces; while Haemmerling (1963) has shown that "if RNA appears at the end of the stalk so do the morphogenetic substances" and that "if RNA disappears the morphogenetic substances also disappear". Although final proof has still to be obtained, Brachet (1963) has concluded that the "morphogenetic substances" must be messenger RNA and it is evident that their synthesis must depend on the activation of special fruiting body genes.

Evidence is also available to show that all the genes on which the production of a mature fruiting body depends are not continually functional but that they are activated in groups in proper sequence (see Haemmerling, 1963). When the fruiting body is close to completion the primary nucleus in the rhizoid undergoes repeated mitosis leading to the production of large numbers of secondary nuclei which are transported up the stalk by protoplasmic streaming. Within the fruiting body they form some 10 000 cysts, each of which is initially uninucleate and within which mitosis and ultimately meiosis occur to produce the haploid gametes. If the fruiting body is cut from the top of a stalk containing mitotically active secondary nuclei, no new fruiting body is formed as would occur in the presence of a primary nucleus. Evidently the secondary nuclei have undergone a change whereby the fruiting body genes have been repressed and the mitosis genes have been activated. It is also clear, since it is not readily reversible, that this new situation is relatively stable.

Just as the fruiting body genes are inactivated once the fruiting body has been completed, so also are the mitosis genes inactivated at the end of the phase of cyst and gamete formation. At this time the few secondary nuclei that are left behind in the stalk show no mitosis although they are fully capable of it. When the mature fruiting body is cut off and an anucleate immature fruiting body is grafted in its place, these secondary nuclei undergo repeated mitosis and normal cyst and gamete formation follows. It therefore appears that a mature fruiting body produces a mitotic inhibitor that diffuses down into the stalk.

The conclusions deriving from all this important work on *Acetabularia* are that the various forms of differentiation are determined by gene-directed syntheses and that during the long and complex process of gamete production there is a sequential activation of groups, or operons, of genes. The final experiments quoted above concerning the reactivation of secondary nuclei for mitosis also show that genes are activated or repressed, as in the bacteria, by cytoplasmic substances or effectors. These conclusions add weight to the belief that in the smaller and more typical unicellular algae and in the protozoans the four main phases of differentiation (vegetative growth, mitosis, encystment and conjugation) may also be controlled, as they evidently are in bacteria, by genetic mechanisms that are activated and inactivated by effector substances.

THE COMPLEXITY OF THE CILIATES

The ciliate protozoans are interesting, first, because in the course of their evolution they have acquired a cell structure which ranks among the most complex known (see Corliss, 1961), and second, because they have been the subject of much important genetical research (see Kimball, 1964). The most elaborate part of the ciliate cell is its cortex, within which lies the mechanism that directs the activity of the field of cilia that commonly covers the whole surface. These cilia were no doubt derived by multiplication from flagella and they share a common morphology. Each cilium is based within the cortex on a kinetosome, or basal body, and the fine structure of this complex in *Paramecium* has been described by Nanney and Rudzinska (1960) and Pitelka and Child (1964). To enable the field of cilia to operate efficiently as a single functional unit the kinetosomes are closely interconnected, and the whole fibrillar system so formed may be connected to a central "motorium" near the buccal cavity. Other such features as the permanent gullet and the contractile vacuoles also add to the cortical complexity, which in turn must demand the presence of a more than usually sophisticated mechanism to allow for proper division during mitosis.

The nuclear mechanism is also surprisingly elaborate, and the impression is gained that a single diploid nucleus may be inadequate for the synthesis of the volume of mRNA needed to support the activities of such a large and complex cell. The minority of ciliates, the "protociliates", contain many nuclei that are all alike, while the majority, the "ciliates proper", typically contain one micronucleus which is diploid and one macronucleus which contains a large number of complete sets of chromosomes. The micronucleus is evidently reserved mainly for conjugation, while the macronucleus is the centre for phenotypic control. During binary fission the micronucleus undergoes typical mitosis, while the macronucleus apparently divides into two by simple constriction or amitosis. During conjugation the micronucleus of each partner undergoes a form of reduction division prior to the exchange of the haploid chromosome complements, while the macronucleus is usually broken down and a new one is later formed from the new micronucleus. Some ciliates, including *Paramecium aurelia*, also show a form of conjugation in which no pairing takes place and in which, after the reduction division, the two haploid nuclei fuse with each other to form a micronucleus which then gives rise to a new macronucleus. This is known as autogamy.

In the contributions of the ciliates to genetics, the genera *Paramecium* (see Beale, 1954, 1964; Sonneborn, 1959), *Tetrahymena* (see Nanney, 1964) and *Stentor* (Tartar, 1961) have been particularly important. It appears that many aspects of cell structure and possibly also of function are not under direct chromosome control and that there exists an elaborate system of extrachromosomal genetics (see p. 46, and also Jinks, 1964; Pitelka and Child, 1964; Sager, 1965, 1966). Most obviously this applies to the synthesis of new kinetosomes and cilia and to their interconnections with other kinetosomes and cilia to form the elaborate pattern of the field of cilia. Most attention has been directed to this problem of extrachromosomal genetics, and therefore the contributions of the ciliates to an understanding of chromosomal genetics, gene control, and differentiation have been relatively small. It is already obvious, however, that mechanisms for selection between the various instructions coded on the chromosomes do exist, and a number of instances of varying complexity have been studied. Certainly by analogy with what has been said above it is reasonable to suggest that particular states of differentiation in ciliates may underlie the phases of cell growth, of cell division, and of conjugation. Sporulation does not seem to occur, perhaps because the elaboration of the cortex makes reorganization for cyst formation impossible.

CELL GROWTH AND DIVISION

Cell growth and division are normally so controlled that in standard conditions a ciliate reaches a standard size at the time when cell division begins. Kimball (1964) has emphasized that some homeostatic mechanism must exist to control cell size, and this also implies that certain conditions must exist within the cell before the preparations for mitosis can begin. The actual control mechanism is probably complex but it seems that it must depend on the activation of some trigger mechanism, which may then activate the special mitosis genes. Cell division in ciliates is particularly complicated since both the micronucleus and the macronucleus must divide. Since the genes of the micronucleus may remain repressed between divisions (Sonneborn, 1954) and since it is only this nucleus that undergoes mitosis, it seems that the trigger mechanism may relate to this nucleus alone.

The macronucleus shows active mRNA synthesis during the interphase, and during cell division its chromosome complement divides amitotically and often unevenly. In *Paramecium aurelia* a gene has been distinguished which promotes uneven macronuclear division (Sonneborn, 1954; Nobili, 1961). Since normal macronuclear size is usually restored before the next cell division, it is clear that macronuclear DNA duplication must continue during the interphase and that some homeostatic mechanism must exist to ensure that the macronucleus attains but does not exceed an appropriate size. It is also evident that any changes in gene expression during the growth phase must involve the macronucleus alone.

Thus, at least in those few ciliates that have been studied in detail, differentiation for cell maintenance and growth involves the selective activation of one nucleus while differentiation for mitosis involves the selective activation of the other. Since both nuclei contain similar chromosomes and both are contained within the same cell, the mechanism of gene control may show unusual complexities. There is some evidence that the different behaviour of the two nuclei may be related to the different cytoplasmic regions in which they lie. The first post-conjugation mitosis separates the future micronucleus from the future macronucleus, and in the case of *Tetrahymena*, Nanney (1953) has suggested that the distinction between them may depend on their locations within the cells. Experimentally he has been able to alter their fate by cytoplasmic disturbance obtained by centrifugation.

An apparently similar example of the effects of the cytoplasm is provided by the fate of the four pronuclei that are formed during reduction division in a *Paramecium* preparing for conjugation. Evidently only the two pronuclei that manage to migrate into what is called the paroral cone region survive, while the other two pronuclei degenerate; Sonneborn (1954) has described a mutant in which all the pronuclei fail to reach the cone region and all degenerate.

CONJUGATION AND AUTOGAMY

Before conjugation or autogamy the macronucleus is usually totally destroyed and afterwards it is reconstituted from the micronucleus.

Evidently, then, this is a second period of cell life that must be dominated by the genes of the micronucleus. The fact that the macronucleus only survives during the period between successive conjugations, or autogamies, again emphasizes the fact that it consists of a set of chromosomes that are created to direct only cell maintenance and growth.

After conjugation, but not after autogamy, there is a period during which no further conjugation can occur. During this period the new macronucleus is formed and two of its gene loci are activated to direct the synthesis of the two "mating-type substances". Kimball (1964) has given evidence for the "successive 'switching on' of first one locus and then the other with the consequence that during an intermediate 'adolescent' period only one of the two substances . . . is found. Thus in this instance there is a temporal sequence of gene expression". The two mating types, or sexes, are each determined by their own particular pair of mating-type substances. When at last both these substances are actively produced the animals are mature, and when two mature animals of different mating types accidentally touch, their cilia adhere together. Thus at least one of the effects of the mating-type substances is to be found in the cell surface.

Although simple in outline the details of this type of sex differentiation are in fact complex. Sonneborn (1960; see also Kimball, 1964), in particular, has studied this phenomenon extensively and he has pointed out that "in nearly all strains, cells that differ in mating type have identical sets of chromosomes and genes". Nevertheless the difference in mating type is due to some difference between their nuclei, and this nuclear differentiation arises in the first nuclei formed after conjugation before the first cell division occurs. Sonneborn has proved that "the differentiation of nuclei in some strains was brought about by the action on one or another of two kinds of cytoplasm present in the cell at the time the nuclei differentiate" and that "the decisive cytoplasmic factors were themselves determined by the action of the kind of differentiated nucleus which had been present in the cell before fertilisation". He further showed that "after a nucleus had once been differentiated by a cytoplasmic factor, its descendant nuclei could not be differentiated in the other direction by later exposure to the other cytoplasmic factor", and therefore that "these nuclear differentiations seem to be irreversible and hereditary".

Sonneborn concludes: "Here for the first time we can answer the question whether nuclear differentiations are the cause or the consequence of cellular differentiation. They are clearly *both*. The cytoplasmic factor—a cellular differentiation—differentiates the nucleus. So nuclear differentiation is a *consequence* of cellular differentiation. But this nuclear differentiation determines two cellular differentiations, one detected as the mating type of the cell and its progeny, the other detected as the cytoplasmic factor which differentiates nuclei. So nuclear differentiation is also a *cause* of cellular differentiation".

IMMOBILIZATION ANTIGENS

In addition to the cell surface modifications that are related to the mating types, it is also known that both in *Paramecium* and *Tetrahymena* the cell surface may continually adapt to external conditions. It is evident that the surface, for instance of *Paramecium*, contains some substance or substances which, when this ciliate is injected into a rabbit, act as antigens and induce the production of antibodies. When one of these antibodies is added to a culture of *Paramecium* possessing the relevant surface antigen, movement ceases and death ensues.

These surface antigens, or immobilization antigens, are of many different kinds, but only one or two are present at a time in the cell surface. The production of each of these antigens is controlled by a particular gene locus, and while one locus is active all the other antigen loci remain inactive. Each locus is activated in response to particular environmental conditions, such as temperature, and when these conditions change the active locus is suppressed and a new locus is activated.

Both Sonneborn (1960) and Kimball (1864) have discussed the selective synthesis of the immobilizing antigens, as well as of the mating-type substances, in terms of enzyme induction and of the Jacob–Monod theory, and they have considered these syntheses to be forms of differentiation. The state of differentiation that is characterized by the production of a particular immobilizing antigen, unlike that associated with particular mating-type substances, is unstable and easily changes, and although the functions of the antigens are not known, it seems probable that each of them may confer some advantage on the cell in the particular circumstances that bring about its synthesis.

DIFFERENTIATION IN THE CILIATES

There clearly exists a considerable amount of evidence, albeit mostly incomplete, to suggest that the ciliates as a group follow the general rules of cellular differentiation whereby gene expression is attuned to the needs of the moment.

Transient phases of differentiation, evidently basically similar to the transient phases of enzyme induction in *Escherichia coli*, occur most obviously in relation to the production of the immobilization antigens, but their existence can also be inferred in connection with cell division, which is initiated by some trigger mechanism that is linked with cell size, and with conjugation, which is at least partly dependent on the production of certain mating-type substances. In all these cases the changes in gene expression, whether they involve the micronucleus or the macronucleus, evidently occur in response to changes in the environment of the chromosomes. Indeed, it seems clear that the whole conjugation cycle, together with the mitotic cycles occurring within it, must depend for its existence on a complex and sequential series of nucleo-cytoplasmic interactions.

The two mating types, or sexes, are particularly interesting since they represent states of differentiation that remain stable during the whole of the inter-conjugation period. Sonneborn (1960) has particularly emphasized this point since "such successive cycles of differentiation of cytoplasm by nucleus, then of nucleus by cytoplasm, may surely be expected to occur also in higher organisms" and since "they could yield precisely the progressive, step-wise cellular differentiations that are so characteristic of the development of higher animals". Already in the ciliates differentiation has evolved from a transient state of gene suppression or activation typical of a bacterium, towards the stable state of gene expression typical of the higher plants and animals, and as more observations come to be made a whole range of intermediates between the two extremes may be discovered among unicellular algae and protozoans.

SOCIAL AND COLONIAL PROTOZOANS

In most groups of free-living unicellular algae and protozoans there has been a tendency towards increase in size and therefore complexity of organization. This has reached its limits for instance in the giant single cells of *Acetabularia*, in the multinucleate and complex cells of the ciliates, and in a variety of multicellular social and colonial forms. The multicellular condition must have been achieved many times and is seen today among flagellates and rhizopods in such extreme forms as the Volvocidae, the Mycetozoa (which is probably not a natural group), and the sponges. Of these the mycetozoans, and especially the myxomycete *Physarum* and the acrasid *Dictyostelium*, have been most intensively studied from the point of view of differentiation.

The mycetozoans are amoeba-like and are characterized by the habit of aggregation to form either multinucleate or multicellular masses. The multinucleate plasmodial forms include *Physarum* while the multicellular forms include *Dictyostelium*, and they have been called social amoebae. There are many genera showing intermediate stages between the solitary and social amoebae. Thus amoebae of the genera *Hart-manella* and *Acanthamoeba* live freely in the soil, but when preparing for encystment they aggregate together. In the closely related soil amoeba *Naegleria* the individuals remain solitary in the presence of adequate bacterial food, become flagellate when excess water is in the soil, and aggregate within a cyst in adverse conditions (Fig. 7; and see

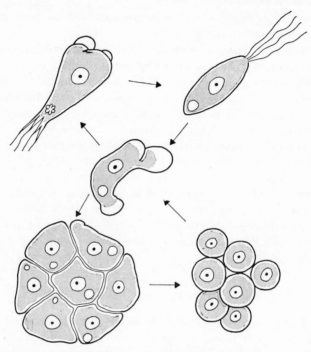

Fig. 7. The various forms of the soil amoeba *Naegleria gruberi*. (Reproduced with permission from Willmer, 1963.)

Willmer, 1963). These various phases of activity, or differentiation, are evidently controlled by trigger mechanisms that are activated by environmental conditions, and it may be presumed that they are medi-ated through changed patterns of gene activity.

THE MYXOMYCETES

In such myxomycetes as *Physarum* the life cycle begins with the release from the spore of a haploid gamete which develops two flagella and conjugates with another gamete. The diploid zygote lacks flagella and moves and feeds in an amoeboid manner. A number of amoebae fuse together in groups to form a small plasmodium, and the nuclei then undergo repeated synchronous mitosis to form a large plasmodium containing many thousand nuclei (Dee, 1962). The synchronous mitotic cycle evidently involves one period of mRNA synthesis for mitosis and another of mRNA synthesis for growth (Rusch *et al.*, 1964). The strictly synchronous nature of the mitotic cycle strongly suggests that the many nuclei are kept in phase with one another because they are all reacting simultaneously to changes in the common cytoplasm, or in other words that the mitosis genes are all activated and inactivated simultaneously by some cytoplasmic effector substance or substances. With shortage of food the protoplasmic mass concentrates into a series of compartments, each of which forms a sporangium with many spores. Each nucleus undergoes meiosis so that each spore is haploid. Finally a resistant cellulose wall is laid down and this has a surface sculptured into a pattern that is typical of the species. Rusch *et al.* (1964) have some evidence that the initiation of sporulation involves the activation of specific genes and the production of specific mRNA.

This sequence of phases of activity from conjugation through mitosis to sporulation must clearly depend on a sequence of phases of enzyme activity, and although the evidence is not yet adequate the implication is that this in turn depends on a sequence of phases of gene activity. It is already apparent that the phases of activity are determined by environmental circumstances, and it is therefore probable that changes in gene activity occur in response to changes, perhaps in the activity of effector substances, in the cytoplasm.

THE ACRASIDS

The evidence from the acrasids, especially *Dictyostelium*, is considerably more extensive. Superficially their phases of activity resemble those of the myxomycetes, but there is no indication that conjugation ever occurs and therefore there is apparently no alternation between haploid and diploid phases. The germinating spore releases a uninucleate amoeba, which feeds and undergoes mitosis to produce a large number of solitary amoebae. If the food supply is then reduced or the amoebae become too crowded, feeding ceases and the animals stream together to form a mass of up to one or two hundred thousand amoebae.

It is now known that this process of aggregation is a complex phenomenon (see Shaffer, 1964). The amoebae are evidently attracted together at one centre by a substance of low molecular weight which they secrete and which has been called acrasin (but see Bonner *et al.*, 1966; other substances including steroid sex hormones also act in this way); they also secrete an extracellular enzyme to destroy acrasin and this evidently helps to maintain a concentration gradient from the centre to the periphery of the aggregation; each aggregation draws amoebae only from a particular territory and in any species the size of this territory is constant and is unaffected by the density of the amoebae (Shaffer, 1961; Bonner and Dodd, 1962); territory size is apparently determined by the secretion from one central cell or group of cells of some substance that inhibits the initiation of any similar aggregations at any point within its diffusion range (Shaffer, 1961, 1963); this or some other substance also acts to slow the rate of movement of all the amoebae within the territory; and finally, under the influence of acrasin the cell surfaces become sticky so that as the cells aggregate they also adhere.

The complexity of this process, which involves specialized secretions and reactions to these secretions, is matched by the complexity of the subsequent events. There is first a phase of migration in which the aggregation crawls like a small slug and leaves behind an extremely thin collapsed slime tube that is evidently secreted by the outermost cells. The anterior cells, or pre-stalk cells, of the slug are clearly distinguishable, for instance by the PAS staining reaction and by immunochemical methods, from the posterior cells, or pre-spore cells (Takenchi, 1963; Gregg, 1965). The slug is attracted towards light and towards heat, and any consequent lowering of relative humidity, even if only slight, is adequate to initiate the formation of the fruiting body.

The front end of the slug then rises into the air at right angles to the substratum (with no relation to gravity), and it is believed that the angle is at least partly determined by the concentration of gas produced by this and any other rising slugs in the close vicinity (Bonner, 1963; Bonner and Dodd, 1962).

As the front end of the slug rises up, the anterior cells, which become vacuolated, pass downwards again, forcing their way through the centre of the mass of the more posterior cells by what has been called "a reverse fountain movement" (Bonner, 1963), until the leading cells touch and stick to the substratum. As more and more of the anterior cells pass downwards in this way a stalk is formed, and on it the previously posterior cells are lifted into the air. Thus two "tissues" are differentiated: the one forming the stalk is composed of vacuolated cells which secrete a stiffening cellulose column; the other forming an apical mass is composed of cells each of which surrounds itself with a cellulose wall and becomes a spore. It is important to note that in this process the stalk cells become so irreversibly specialized that, like the differentiated tissues of the higher plants and animals, they are doomed to die.

DIFFERENTIATION IN SOCIAL AMOEBAE

It is obvious that in the social amoebae there must exist a variety of inducer and repressor substances which control the various phases of the life cycle from the germinating spore to the formation of the fruiting body. The most important evidence in support of this comes from the results of experiments with *Dictyostelium* which have demonstrated: the existence of acrasin and probably also other effector substances; the manner in which, when a migrating slug is cut in two, each half regenerates the missing part so that differentiation already begun in the pre-stalk cells and in the pre-spore cells can be reversed when circumstances change; and the way in which, over an extremely wide size range, stalk cells and spore cells are formed in the proper proportions.

The obvious suggestion is that all such aspects of differentiation are the outcome of effector-dominated changes in the pattern of gene activity. However, this conclusion has been questioned, especially by Wright (1963, 1964, 1966), who has considered whether during differentiation, when the activities of certain enzymes are enhanced, this could be due basically to the availability of new substrates or to the release of the enzymes from some previously inactive form. She has suggested that the morphogenesis of the fruiting body in acrasid aggregations may be uniquely simple, "that all the 'informational RNA' necessary for the synthesis of the critical enzymes controlling morphogenesis exists in the (free-living) amoebae, and that the control mechanisms for biochemical differentiation therefore depend entirely on the accumulation and activation of enzymes and on sequential substrate availability". Certainly there is good evidence for the importance of substrate control *in vivo* in the absence of changing enzyme levels (Wright, 1964), and since the aggregating cells do not feed it is possible that substrate levels in acrasid aggregations may change progressively.

There is no conclusive evidence either for or against this point of view, but Bonner (1963) has strongly argued for the contrary conclusion that the sequence of changes in the formation of a *Dictyostelium* fruiting body are more probably gene-dominated. Certainly the process itself is gene-controlled since mutants have been found which are unable to aggregate. Bonner concludes that since a whole colony can be obtained from one spore, this spore "is the repository of all the information necessary to complete the development . . . presumably in the genes of the single nucleus", and that "to repeat the obvious the instructions these genes set forth are completely dependent upon the environment. This includes not only the immediate environment of the cytoplasm enclosing the nucleus, but the environment around the cells as the result of the actions of the other cells, and finally the environment at large, the temperature, the humidity and so forth".

This point of view is supported by the fact that the stalk of the acrasid fruiting body contains cells that are so highly differentiated that they cannot revert to the amoeboid state, which may indicate, as in higher organisms, that certain aspects of gene expression have been so firmly repressed that never again can they be activated.

EXTRACHROMOSOMAL GENES

For more than fifty years examples of extrachromosomal inheritance in unicellular algae and protozoans have been accumulating, and recently these have been well reviewed, for instance by Jinks (1964), Wilkie (1964) and Sager (1965, 1966). The evidence concerns especially the two centrioles which function as opposite poles for the spindle apparatus, the kinetosomes which underlie flagella and cilia, the plastids of photosynthetic plants, and the mitochondria which are energy-transforming organelles. It has been suggested that all these are self-replicating structures which contain within themselves the coded instructions for self-synthesis. If this is true the total hereditary material of any eukaryote cell must include many extrachromosomal genes, which could be composed of DNA, RNA or even some other type of coding molecule, and the question arises whether these may make a significant contribution to any phase of cell differentiation. Such extrachromosomal genes, often called plasmagenes, are collectively referred to as the plasmon to distinguish them from the chromosomal genes which

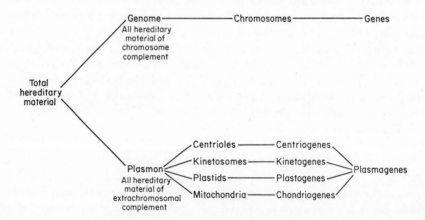

FIG. 8. The main subdivisions of the hereditary material of the cell. (Reproduced with permission from Jinks, 1964.)

collectively form the genome. A classification of genetic material in these terms is shown in Fig. 8.

In addition, a variety of extrachromosomal genetic elements have been described, especially in *Paramecium*, that have proved to be intracellular viral or bacterial symbiotic organisms (see Sonneborn, 1959; Beale, 1964). These organisms are inherited in the cytoplasm in the manner of congenital diseases and they may induce morphological or physiological changes in the host cells. Such changes may be regarded as examples of pathological differentiation.

CENTRIOLES

Two of these organelles are evidently present in all animal cells, although their existence in plant cells is still in doubt. Before mitosis a small particle appears alongside each centriole, and during mitosis this develops into a new centriole. Each new cell then inherits one old and one new centriole. Thus the centrioles have a visible continuous existence from one animal cell generation to the next, and since they form "the only residual protein found in the sperm heads of some animals" (Jinks, 1964) they may also maintain a continuous existence from one metazoan generation to the next. It is important to emphasize that centriole duplication does not depend on centriole division. The new centriole grows alongside the old one and at right angles to it, and it is mainly for this reason that it has been supposed that the old centriole must contain "centriogenes" with the necessary coded information (see Mazia, 1961). However, these centriogenes remain hypothetical, no genetic material has been identified in any centriole, and so far no experiment has shown whether or not a cell deprived of its centrioles can or cannot make new ones.

Also, even if it is true that a centriole is a self-coding self-replicating structure, its replication is evidently still under the ultimate control of the chromosomal genes. Certainly it is probable that the message to commence replication must originate in the "mitosis operon" and pass to the centrioles, perhaps in the form of an effector molecule. Thus part of the process of differentiation for mitosis may involve the activation of certain plasmagenes.

KINETOSOMES

In the protozoans there is clear evidence of a close relationship between centrioles and the kinetosomes (basal bodies or blepharoplasts) that underlie flagella and cilia. Structurally these two organelles are closely similar (see Grimstone, 1961), and in the flagellates, but not the ciliates, they are closely connected. Thus in *Barbulanympha* each newly-formed centriole gives rise to a new kinetosome and thus to its associated flagellum, so that in each new cell after division one centriole-kinetosome-flagellum complex is new and the other is old. Indeed, it has been suggested that during evolution the first kinetosomes arose as derivations of centrioles.

It has also been suggested that kinetosomes are autonomous, and that in the extreme case of the ciliates both the production of new kinetosomes and the formation of the specific pattern of the kinetosome system are specified by genetic material that is stored within this system (see Lwoff, 1950; Sonneborn, 1963). It has been shown both in *Stentor* and *Paramecium* that during the regeneration of cell fragments the formation of a new cortex is dependent on the survival of at least part of the old cortex (see Jinks, 1964), and that the genetic information in the chromosomes is inadequate for this purpose. There is also good evidence for the presence of active DNA in the kinetosomes of the flagellate *Trypanosoma* (Steinert *et al.*, 1958) and of the ciliate *Tetrahymena* (Seaman, 1960, 1962). However, doubts that the evidence for the existence of kinetogenes is indeed adequate have been expressed by Kimball (1964) and Pitelka and Child (1964). Certainly the evidence is inadequate to support any conclusion that the phases of differentiation shown by the protozoans may depend on the periodic activation of kinetogenes, although it can be suggested, for instance, that the growth of flagella in wet conditions by the soil amoeba *Naegleria* (Willmer, 1963) may involve activation of this type.

PLASTIDS

The chloroplasts of plant cells have long been suspected of possessing some genetic autonomy. Recently Schiff and Epstein (1965) have concluded that experiments with *Euglena* and other species "seem to leave no doubt that a DNA exists in the chloroplast" and that there is reason to believe "the information for the construction of the *Euglena* chloroplast or proplastid resides in the organelle itself" (see also Granick, 1963; Sager, 1965). However, it is also clear from experiments with maize and *Oenothera*, that close connections must exist between the nucleus and the chloroplast so that "at the very least, the replication of organelles such as the chloroplast is responsive to overall signals for cell division which coordinate the division of the cytoplasmic organelles, nucleus, and of the cell itself" (Schiff and Epstein, 1965). So once again there is evidence that the activation of plasmagenes may be an integral part of the process of differentiation for mitosis. It is even possible that at least part of the "overall signal" may take the form of DNA polymerase, which might act both on the DNA of the chromosomes and on that of the chloroplasts.

There is also an interesting suggestion that the chloroplasts of eukaryote plant cells may have originated in some early symbiotic relationship between eukaryote animal cells and photosynthetic bacterial or blue-green algal cells. Certainly there are close similarities between chloroplasts and modern unicellular blue-green algal cells (see Echlin, 1966), and Hall and Claus (1963) have described the case of a blue-green alga which lives within the flagellate protozoan *Cyanophora paradoxa*, has lost its cell wall, and divides in step with the host cell. If all chloroplasts have indeed originated in this way it would not be surprising to find that they contain the DNA code needed to specify their own replication.

MITOCHONDRIA

These organelles share with plastids a structure based on membranes, a function concerned with electron transfer and phosphorylation, and the possession of DNA (Sager, 1965; Schiff and Epstein, 1965). The case for the genetic autonomy and continuity of mitochondria is supported mainly by evidence from fungi, especially yeast and *Neurospora*, in which a number of apparently extrachromosomal mutants leading to respiratory deficiency have been detected. There is also inconclusive cytological and biochemical evidence to support the view that these mutations lie within the mitochondria and that they involve certain chondriogenes that specify respiratory enzymes. Although these findings refer only to fungi, the ubiquity and basic similarity of mitochondria in all eukaryote cells suggest that they may be generally applicable. From a detailed review Wilkie (1964) concludes that "the evidence from the genetic and biochemical analysis of respiratory deficiency in yeast is heavily weighted in favour of cytoplasmic inheritance *in sensu stricto*." However, he also shows that mitochondrial synthesis most probably depends on the joint action of chromosomal genes and of chondriogenes; neither can apparently act alone.

Although much further information is needed regarding mitochondria and all other cell organelles, it is already possible to question, as Sonneborn (1963) has done, whether the chromosomal genes ever contain enough information to construct a whole cell or whether one cell must always pre-exist to provide the basic pattern according to which a new cell is constructed. This pattern may be found partly within specific organelles and partly in the more general form of the whole molecular organization of the cytoplasm. This question is discussed on p. 55.

Conclusions

The eukaryote cell is the dominant modern cell type, and in contrast to the prokaryote cell it is characterized by a much larger size, by a more elaborate structure which includes a variety of organelles, by a much larger genome, and by diploidy in some phase of the cell cycle. One recurrent evolutionary theme shown by these cells has been a tendency towards multicellularity, which has led to the appearance of all the higher groups of organisms.

THE ORGANIZATION OF THE GENOME

The most important single advance shown by the eukaryote cell of the unicellular algae and the protozoans has been the greatly increased genome, which has at least ten times the informational content of the prokaryote genome. This has enabled the eukaryote cell to develop its greater size and complexity as well as its greater versatility in the face of a hostile world. More important still it has endowed the cell with much greater potentialities for further evolution, especially towards the more elaborate multicellular condition.

The problem of packaging this large genome has been solved by the evolution of a chromosome with a histone backbone on which the DNA is stabilized by reversible ionic bonding, while the problem of the efficient handling of the genome during cell division has been solved by the evolution of the complex mitotic apparatus. The manner in which the DNA thread is folded on to the histone (which may itself also be folded) is quite unknown and is a matter only for speculation (see DuPraw, 1966). Of more importance is the manner in which the DNA-histone binding appears to be broken locally whenever certain genes need to be transcribed. Indeed, it has often been suggested that it is a primary function of the histones, by binding or unbinding with appropriate operons, to control the pattern of gene activity and thus the state of differentiation of the cell (see p. 88). While it remains probable that only unbound operons are active, it is becoming increasingly improbable that the histones play more than a passive role in gene activation and repression (Johns and Butler, 1964). It will be recalled that an efficient method of gene control was already established in the histone-less chromosomes of the prokaryote cells, and it seems probable that a similar method also exists in the eukaryote cell. If this is true then the presence of histone merely poses an extra problem which is solved by the detachment of the activated genes. In other words gene activation and repression in the eukaryote cell may depend, as in the prokaryote cell, on gene-specified repressors which in addition must act directly or indirectly on the DNA-histone binding (for further discussion see p. 88).

Another important advance in the eukaryote genome is the common condition of diploidy. One set of chromosomes comes from each of the two parent cells, either by conjugation in unicellular organisms or by fertilization in multicellular organisms. This, together with the fact that recessive genes may be hidden in large numbers within the population, and that gene recombinations commonly occur, vastly increases the amount of variation that may be expressed within a species. Each generation of exconjugants, or zygotes, may include a novel range of variations of which some may possess selective advantages, and thus a species may change slowly and progressively, particularly in changing environmental conditions.

This eukaryote condition of diploidy has depended on the evolution of another mechanism for handling the chromosomal DNA, the mechanism of meiosis by which the haploid condition is reached before conjugation, or fertilization. Meiosis appears in fact to be a variant of mitosis and the two may perhaps share a common evolutionary origin.

CONJUGATION AND THE EVOLUTION OF SEX

It is possible that conjugation in diploid eukaryotes may have evolved from the type of conjugation found in haploid prokaryotes, which was itself pathological in origin. If this is so, then what was originally viral DNA has become fully incorporated into the cell genome and has lost its identity, while the cell, by becoming diploid, has for the first time derived the full advantages of conjugation.

Conjugation is not, of course, a form of reproduction in unicellular organisms, all of which rely on cell division. Its main function is to promote gene spread and recombination, but it may also have another consequence which can be illustrated by reference to *Paramecium*. A single individual multiplying by mitosis in ideal conditions gives rise to a large number of individuals which are collectively called a clone. It is a common observation, at least in some species, that with the passage of time such a multiplying clone becomes senescent and may ultimately die out. If, however, individuals are allowed to conjugate with members of another clone, the exconjugants, besides acquiring new gene mixtures, also acquire a new lease of life so that senescence and death are postponed.

This is a strikingly similar situation to that in a multicellular plant or animal in which growth by mitosis produces a clone-like mature body which is also doomed to senescence and death. In this case conjugation, now called fertilization, is confined to the germ cells, and the zygotes so produced, besides acquiring new gene mixtures, also acquire a new lease of life.

It certainly seems probable that the process of sexual reproduction in the higher organisms must have evolved directly from the system of non-reproductive conjugation of the unicellular organisms, although the intermediate steps are unknown. An interesting possible sequence of stages, which themselves certainly have no evolutionary significance, are provided, first, by *Paramecium* in which the male pronuclei move without cytoplasm into the other conjugant; second, by various rumen ciliates of the families Cyclopostliidae and Ophryoscolecidae in which the male pronuclei are surrounded by cytoplasm and a cell wall and are each provided with a single flagellum to propel them on their journey (Calkins, 1933); and third, by the ciliated and partly cellularized

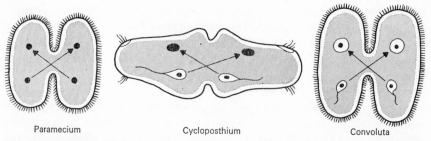

Paramecium   Cycloposthium   Convoluta

FIG. 9. Conjugation in ciliates and sexual reproduction in metazoans. In *Paramecium* and *Cycloposthium* (which has sperm-like male pronuclei) conjugation leads only to gene recombination, while in the metazoans, like the platyhelminth *Convoluta*, it leads also to egg production.

acoelan flatworms (Platyhelminthes) in which the female pronuclei are also surrounded by cytoplasm and a cell wall and are therefore termed ova (see Fig. 9). In the ciliates each conjugating individual is rejuvenated by the fusion of the pronuclei, but in the acoelan flatworms the rejuvenation process, confined within the cell wall of the ovum, is limited to the next generation. Also in the acoelans each individual produces not one but a considerable number of ova and spermatozoa, so that the process is transformed from conjugation into reproduction.

The habit of exchanging haploid pronuclei is certainly one of the most important contributions made by the unicellular organisms to the multicellular organisms. Apart from the value, stressed above, of permitting gene spread and recombination in the population, it frees the multicellular organisms from the necessity of reproducing by fission or budding. Although this type of asexual reproduction does occur in plants and in the lower animal phyla, it would clearly have restricted the evolutionary potential of the higher animals, which would certainly have been different from, and probably simpler than, those that now exist.

AGEING AND REJUVENATION

Unfortunately there is not yet any widely accepted theory to account for the progressive ageing of cells and of organisms (see Curtis, 1963, 1966), or for the rejuvenation that follows conjugation or fertilization. It is clear, however, that senescence and ultimate death are the certain fate of any cells which develop that stable form of differentiation seen, for instance, in the stalk of the fruiting body of *Dictyostelium* and in the tissues of an adult mammal. In all such cases it appears that the majority of the genes have become firmly and perhaps irreversibly repressed and that the cell is operating with ever-decreasing efficiency on the few genes that remain active (see also p. 120).

In many of the higher plants, in which tissue differentiation is not so firmly established, vegetative propagation may continue apparently indefinitely. In such cases, as when fragments of ageing leaves give rise to complete and rejuvenated plants, it is necessary to assume that there must have been a reactivation of the entire genome which thus has become as totipotent as it is in a zygote.

It is possible that the situation in unicellular organisms is similar. Thus ageing clones of *Paramecium* seem to show a progressive inability to synthesize essential enzymes even in ideal environmental conditions, and this may well be the result of ever-decreasing gene activity. Conversely, after conjugation, enzyme production is active once more, and it is possible that such rejuvenation is associated with the reactivation of all the genes.

Although these observations offer no explanations for the twin problems of ageing and rejuvenation, they do suggest that in all types of eukaryote cells, whether in unicellular or multicellular organisms, ageing is accompanied by progressive gene closure and rejuvenation by the sudden activation of previously quiescent genes, and to this extent both ageing and rejuvenation may be regarded as special facets of the phenomenon of differentiation (for further discussion see p. 119).

UNSTABLE AND STABLE DIFFERENTIATION

Relatively little is known in the unicellular algae and protozoans concerning that type of unstable differentiation which in bacteria has been analysed by Jacob and Monod in terms of rapidly reversible gene-controlled enzyme induction. However, that such transient types of differentiation do exist is clearly shown, for instance, in *Paramecium* in the case of the effector-induced, gene-controlled syntheses of cell-surface macromolecules that have been recognized as the "immobilizing antigens". It seems probable that the syntheses of these substances, which are rapidly promoted or repressed in response to environmental changes, may be controlled by simple effector-repressor-operator units of differentiation of the Jacob–Monod type. More elaborate phases of differentiation are also clearly distinguishable, as they also are in bacteria, and these lead, for instance, to mitosis, encystment, the different phases of polymorphism, or to conjugation. All these may be regarded as unstable forms of differentiation since they all ultimately give place to other types of cell activity.

More stable forms of differentiation are also found in the unicellular eukaryotes, and one example is provided by the differentiation of sex. The manner in which this is achieved varies in detail in different organisms. In *Paramecium*, for example, sex remains stable in the individuals of one clone throughout the whole of the inter-conjugation period, which is analagous to the situation in many metazoans in which sex is commonly stable throughout the whole of the inter-fertilization period. It has already been suggested that a clone may be at least roughly equated with the body of a higher plant or animal. However, in *Paramecium* the stability of sex is known to be determined at con-jugation by the action of cytoplasmic factors which activate or re-press certain universally present genes, while in the higher metazoans the stability of sex is commonly determined by the presence or absence of particular groups of genes.

Still more stable differentiation, as stable indeed as that found in the highest plants and animals, also occurs in a few unicellular eukaryotes. The most extreme case so far described is that of the stalk cells of the fruiting body of the cellular slime mould *Dictyostelium*. These cells are so irreversibly differentiated for their role in supporting the spore mass that they are doomed to die.

Stable differentiation is clearly something that becomes necessary with increasing complexity of structure, since this involves the evolution of groups of cells that are specialized to promote the interests of the other more generalized cells. Indeed, the fruiting body of *Dictyostelium*, which has evolved to serve the needs of the spores, is directly compar-able to the body of a higher plant or animal, which is itself a fruiting body that has evolved to serve the needs of the gametes. Stable tissues, whether of unicellular or multicellular origin, that exist to carry, protect, and nurture the reproductive cells can be regarded as merely temporary and dispensable wrappings for these cells. In such stable tissues only those genes remain active which are involved in the maintenance of the tissue form and function, and in all cases the evidence emphasizes that death is the price ultimately paid for such an extreme form of differentiation.

DIFFERENTIATION AND EXTRACHROMOSOMAL INHERITANCE

There is, however, another and even more obvious form of stable differentiation in unicellular algae and protozoans, which is different in nature and which does not lead to the death of the cells. It involves the expression of those characteristics by which a species is classified and which are commonly situated in the outer part, or cortex, of the cell. Although controversy still continues, it now seems probable that much of the stable structure of the cortex may be specified not by the chromosomal genes but by some form of coded information in the cortex itself. In particular it is possible that both centrioles and kinetosomes may be self-coding, using either RNA or DNA. Regarding RNA, it has been suggested by Sager (1965) that this may be the most primitive form of coding material, that life in the genetic sense may "have begun with an RNA-protein complex which used the environment as a source of building blocks for further synthesis of itself", and that the DNA coding mechanism may have been a later refinement.

The existence of extrachromosomal genetic information is most firmly established in relation to plastids and mitochondria (Wilkie, 1964; Sager, 1966), this conclusion being based on evidence of extrachromosomal mutation, on the presence of DNA in these organelles, and on the assumption that all DNA has a genetic function. It has also been suggested that both plastids and mitochondria may have originated from symbiotic organisms, probably prokaryotes, and that is why they possess their own self-specifying DNA code.

Certainly the possession of extrachromosomal plasmagenes still further increases the potentialities of the cell, and it is probably the existence of these cytoplasmic genes that has permitted the development of the extreme condition of complexity reached by the ciliate cortex. The determination of body form in this way must, of course, be limited to organisms that reproduce by cell division in such a way that each daughter cell receives a large part of the parental cytoplasm with its enclosed plasmagenes. The body form of multicellular plants and animals must usually be created *de novo* in each generation, and its details must therefore be coded in the chromosomal genes. However, in both unicellular and multicellular organisms the plastids and mitochondria, or their plasmagenes, have a continuous existence from generation to generation, although in multicellular organisms their inheritance tends to be mainly maternal.

In conclusion it must be stressed that even if certain aspects of the differentiation of the eukaryote cell are under the control of plasmagenes, these plasmagenes may themselves be activated or repressed in response to signals from the chromosomal genes. In particular it must be noted that self-coding organelles always multiply approximately in step with cell multiplication and that they are inactivated in spores or cysts. Thus the hierarchy of regulator genes on which differentiation depends may control genetic events both inside and outside the nucleus.

CHAPTER 4

Tissue and Organ Formation

There are three major groups of multicellular organisms that have survived to the present day: the Fungi, which also include certain unicellular organisms; the large group of plants which could be collectively called the Metaphyta and which include the multicellular algae, bryophytes, gymnosperms, and angiosperms; and the Metazoa. It seems certain that these represent at least three quite separate essays in multicellularity, but from the point of view of their control of differentiation they are evidently all essentially alike.

In the case of the fungi an introduction to the role of the chromosomal genes has been given by Fincham and Day (1963), while the relation of extrachromosomal factors to differentiation has been considered by Srb (1963). In the metaphytes most attention has been paid to the angiosperms in which, unlike the higher animals, the embryonic type of tissue differentiation is not confined to an early phase of growth but continues throughout the adult phase. This is because of the persistence of embryonic areas at the apices of both shoot (Erickson, 1959; Allsopp,

c

1964) and root (Brown, 1963), and it appears that the newly-formed cells differentiate partly according to the position they come to occupy and partly according to the state of the plant at the moment. The positional effects give rise to the characteristic patterns of branches, leaves and roots; the state of the plant may vary hormonally, for instance in relation to the light (Vince, 1964), and this may determine, among other things, whether the apical cells become quiescent or whether they differentiate as flowers instead of leaves, or whether the roots develop tubers (Miksche, 1964; El-Antably and Wareing, 1966). It appears that both the positional and the hormonal effects are mediated in the usual way by the action on the genome of effector molecules present in the cytoplasm. In the words of Steward and Ram (1961), who have made a general review of the whole problem, the manner in which plant tissues differentiate "becomes a problem of chemical growth regulation" and involves "an array of regulatory substances (which) intervene to control or modulate these events within limits that are set by the genetic machinery".

The higher plants are structurally strikingly different from the higher animals, but their methods of differentiation are nevertheless strikingly similar (see Raven, 1958; Grobstein, 1959; Brachet, 1960; Markert, 1965). The tissues of animals are also formed in the correct pattern in response to what are evidently positional effects, and the actions of various hormones on the differentiation and growth of tissues are well known. Thus the basic strategy of gene control is the same in all types of multicellular organisms, and the words of Steward and Ram, quoted above, can be applied equally to animals.

One apparent difference between plants and animals that is often emphasized is the common reversability of differentiation in the higher plants and the evident irreversability of differentiation in the higher animals. However, it has now been shown that the nucleus from a fully differentiated intestinal cell of a *Xenopus* tadpole, when introduced into an enucleate *Xenopus* egg, can support full development into a sexually mature animal (Gurdon and Uehlinger, 1966). Thus the difference in the stability of differentiation between plants and animals may be more apparent than real and the only practical problem may concern the methods required to reopen and reactivate the genome.

Since there are few, if any, differences in principle between the mechanisms of differentiation in various groups of multicellular organisms, it is proposed to concentrate mainly on the metazoans and to consider first the question of embryonic organogenesis.

EMBRYONIC DIFFERENTIATION

In the metazoans the most obvious and dramatic phase of differentiation is that in which the body is moulded and the tissues and organs are formed, and this is a phase that occurs mainly, though not entirely, in the embryo. Embryonic differentiation is a process which must have been established in all its essential features in the earliest multicellular animals and which, it is reasonable to suppose, must have evolved through the exploitation of some process that already existed in the ancestral unicellular organisms.

In its most extreme forms, as seen for instance in vertebrates, it involves an extraordinarily complex pattern of changes, including cell movement, cell division and cell differentiation (see Abercrombie, 1965), which must in turn be dependent on a multitude of stimuli and of reactions occurring in the correct positions and in the correct sequence. Knowledge of the complexity of this pattern is derived mainly from descriptive accounts of embryonic tissue and organ formation, and far less is known of the multitude of control mechanisms that must exist to direct it. The only practical approach to such complexity is to resolve the problem into its simplest possible elements and to concentrate on the analysis of what appear to be single steps. From the limited number of such analyses that have been made the essential nature of embryonic differentiation is already beginning to emerge. In particular it appears that each single step in tissue differentiation depends on the trigger action of some particular inducing substance, or substances, on cells that are competent to respond.

Some years ago great efforts were made to discover the chemical nature of these inducing agents, and especially of the agent produced by the dorsal lip of the amphibian blastopore (see Needham, 1942). It was evidently felt that an understanding of inducer chemistry could provide a significant insight into the process of differentiation itself. There was at the time a similar belief among endocrinologists in relation to the chemistry of hormones, and among pathologists in relation to the chemistry of carcinogens. It is now being increasingly appreciated that inducing agents, like hormones and carcinogens, are merely triggers and that the important part of the reaction is that which occurs within the responding cell. Deuchar (1966) has commented that in this work the cart has been put before the horse since "it is usual in biochemical work to know the course of a reaction before going on to investigate the conditions that control it". It is also abundantly clear that the action of a trigger is not very specific, and a long list of unnatural agents that can cause induction is given by Needham (1942). Nevertheless it is regrettable that the chemical nature of the natural inducing substances remains so obscure.

THE IMPORTANCE OF THE EGG CYTOPLASM

The most important information deriving from more than half a century of experimental research into the earliest phases of embryonic differentiation has been excellently reviewed by Saxén and Toivonen (1962) and Deuchar (1966), and here it is only necessary to outline the main points that relate to the present argument. Some of the most significant results have come from studies of the particularly large eggs of the amphibians.

In the first place it has become evident that early embryonic differentiation depends, at least in part, on events occurring in the primary oocyte while it is still in the ovary. The amphibian oocyte, while accumulating its yolk reserves, may increase in volume more than a million and a half times, and this is achieved not by the passive acceptance of preformed macromolecules but by the absorption of smaller molecules which are then used in the synthesis of the larger molecules. Experiments with primary oocytes using radioactive tracers have shown active synthesis of RNA on the chromosomes and of enzymes in the cytoplasm. It is also well known that the long thin chromosomes are lined by large numbers of paired loops of despiralized DNA and that these loops are the actual sites of RNA synthesis (see Fig. 10). It is even possible that each loop may represent a single active gene.

All this activity leads in the fully grown primary oocyte to great accumulations of a variety of macromolecules, including RNA molecules, which are not evenly distributed throughout the cytoplasm but which are zoned to form a three-dimensional pattern of gradients. There are also gradients in the density of the mitochondria, which are evidently created through the activity of extrachromosomal DNA, and which in the past have been described in terms of gradients of respiratory activity. In the words of Fischberg and Blackler (1961): "These gradients run parallel to the egg axis and are radially symmetrical about it. As a consequence all the meridional slices going from the upper pole of the egg to the lower one (like segments of an orange) contain the same substances distributed in the same way". This symmetry determines the future antero-posterior axis of the embryo.

In amphibians the dorso-ventral axis is not normally decided until after fertilization, when the point of entry of the sperm fixes the mid-ventral line. The change in symmetry from radial to bilateral is accompanied by changes in distribution of the cytoplasmic components, particularly in the outermost layer, or cortex, of the egg. The most obvious of these cortical changes in the frog egg involves the appearance of a grey crescent along the junction of the pigmented with the non-pigmented areas; the centre of this crescent lies opposite to the point of entry of the sperm. The importance of the cortex in early embryonic

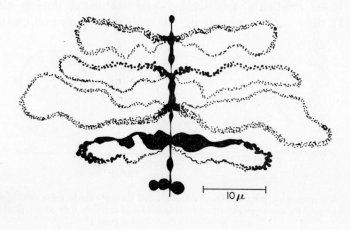

FIG. 10. Above, a segment of the lampbrush chromosome of a newt oocyte, and below, the suggested structure of a pair of lampbrush loops. (Reproduced with permission from Gall, 1963.)

development was first stressed by Dalcq and Pasteels (1937) and Pasteels (1940, 1941), who concluded that the position of the dorsal lip of the blastopore, and subsequently of the nervous system, is determined by the interaction of the two morphogenetic gradients, the one within the yolk and the other in the cortex. More recently Curtis (1960, 1962) has reopened the question by developing a technique for grafting cortical material from one first-cleavage embryo to another. In particular he has stressed the importance of the cortex from the grey crescent area in determining the position of the blastopore and he has concluded "that the cortical material definitely possesses morphogenetic properties which may be transferred with it".

There thus appear to be two main chemical gradients within the fertilized amphibian egg, the one determining the antero-posterior axis being mainly endoplasmic and the other determining the dorso-ventral axis being mainly cortical. The endoplasmic gradient is easily disturbed by centrifugation, but the cortical gradient lies within the stiffer ectoplasm and is not easily disturbed. The general belief is that when these gradients are divided and stabilized during cleavage (Fig. 11), the result is a variety of cells each with its own unique type of

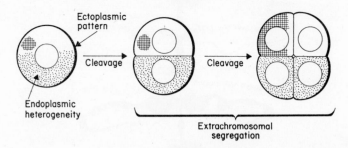

FIG. 11. A diagram showing how four different cell lineages could arise during the cleavage of a zygote by the segregation of ectoplasmic and endoplasmic zones. (Reproduced with permission from Jinks, 1964.)

cytoplasm and each ultimately giving rise to a particular region of the embryo. If such a process of segregation of fragments of the cytoplasmic gradients is the essence of the earliest steps in differentiation, then in order that all embryos shall develop alike it is clearly necessary for the planes of cleavage to be precisely determined relative to these gradients. It is well known that at least the first few cleavage planes are indeed precisely oriented, and since the plane of each division is related to the position of the centriole, it seems that this organelle may play a critical role in early embryonic development.

As is well known, the importance to later differentiation of the type of cytoplasm contained in the cells of blastulae and gastrulae seems to vary from high in the so-called mosaic eggs, for instance of the ascidians, to low in the so-called regulative eggs, for instance of the vertebrates. In a mosaic embryo the importance of the cytoplasm can be demonstrated by transposing in the blastula small groups of future ectodermal and endodermal cells; the tendency then is for the cells to follow the dictation of their own cytoplasm rather than that of the position in which they find themselves. With a regulative embryo, on the other hand, the tendency is to follow the dictation of the position, which presumably means the dictation of the chemical characteristics of the micro-environment of that position. However, there may be little, if any, essential difference between these two types of development: in the mosaic embryo the inducing agents associated with a particular type of cytoplasm evidently tend to remain within the cell, while in the regulative embryo the inducing agents evidently tend to escape to create a characteristic local environment. The range of intermediates between the mosaic and regulative extremes, which runs for instance through molluscs, echinoderms and coelenterates, may reflect little more than differing degrees of cell wall permeability to the inducing agents.

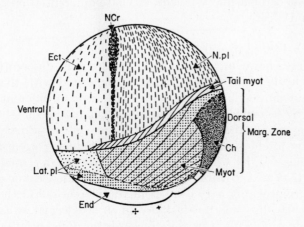

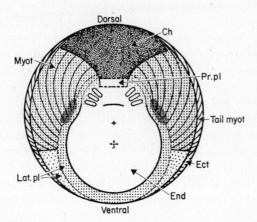

FIG. 12. A newt blastula seen from the left side (above) and the vegetal pole (below) to show which cells will give rise to the various tissue and organ systems. *Ect,* ectoderm; *end,* endoderm; *ch,* notochord; *lat. pl,* lateral plate; *myot,* myotomes; *n cr,* neural crest; *n. pl,* neural plate; *pr. pl,* prechordal plate. (Reproduced with permission from Saxén and Toivonen, 1962.)

The important conclusions are: that the earliest steps in cell differentiation leading to the positioning of the major organ systems and thus to the establishment of the major body axes are related to chemical gradients in the cytoplasm; that these gradients are cut into segments and so stabilized during cleavage; that consequently in the blastula each cell has its own particular type of cytoplasm; that in the blastula it is already possible to predict which cells will normally give rise to which tissue and organ systems (see Fig. 12); and that it is reasonable to suggest that cytoplasmic type and ultimate differentiation are related as cause and effect.

SECONDARY INDUCING AGENTS

It is clear, however, that if the first steps in differentiation are promoted by inducing agents which are inherited with the chemical gradients in the fertilized egg, the second steps in differentiation are promoted by secondary inducing agents which are presumably synthesized in response to the trigger actions of some of the primary inducing agents. The production of these secondary inducing agents is active after gastrulation, by which time there already exists a small range of cell types that are recognizable, for instance as ectoderm or endoderm, and that are probably only competent to differentiate further in their own particular direction, for instance as an ectodermal or as an endodermal derivative. From this point onwards the differentiation of tissue and organ systems depends on the production of an increasingly elaborate pattern of secondary inducing agents, which act only upon those cells in the immediate vicinity that are competent to respond. Thus as differentiation becomes more and more specific the fate of any particular cell depends on a combination of its previous history, which determines its competence, and of the influence of whatever relevant inducing agents are produced in its neighbourhood.

In this process cell movement plays a critically important role. It is often only through such movement that groups or sheets of cells of different tissue systems come to lie so closely adjacent that they can influence each other's further differentiation by a mutual exchange of inducing agents. Thus organogenesis also depends on a pattern of cell movements which must be undertaken at the appropriate time and place, no doubt in response to yet another series of particular chemical signals.

The first discovered and best known of the secondary inducing agents is that which diffuses out from the region of the dorsal lip of the amphibian blastopore, the cells of which form the "primary organizer". The discovery of this active region of the gastrula was the great contribution of Spemann (1901, 1938) to experimental biology, and he visualized embryonic differentiation and development as depending on the activities of a whole succession of such organizer regions, each beginning its task of induction at the point where the previous induction had ceased. The first indication that such inductions are dependent on chemical agents and not on living cells was when Holtfreter (1933) obtained typical inductions with dead primary organizer, although later information has suggested that this evidence is less critical than it appeared at the time, since almost any treatment that causes some cytolysis may result in induction (Barth, 1941; Holtfreter, 1947).

As has already been mentioned, nothing is known as yet of the chemical nature of these embryonic inducing agents. However, evidence for the existence of different chemical agents that are capable of different specific inductions has come from the researches of Yamada and Toivonen (see review by Deuchar, 1966). They have extracted from various adult mammalian tissues a substance, or substances, that in amphibians can cause mesoderm induction in suitably competent embryonic cells, as well as another substance, or substances, with different chemical properties that can cause neural induction. It is impossible to decide at the moment whether such substances taken from adult mammals bear any relation to the normal embryonic inducing agents of the amphibians, but at least the results do indicate that different inductions may depend on different inducing agents. Saxén and Toivonen (1962) have shown great confidence in this evidence and have also concluded that these inducing agents "are obviously large-molecular protein-like factors".

Another well known, and historic, example of tissue induction in amphibians is provided by the correlated development of the lens and eye cup (Spemann, 1901, 1912). The eye cup forms as an evagination of the brain and it then acts as an organizer to induce lens formation in the overlying epidermis. When an eye cup is grafted into some abnormal position the overlying epidermis is equally capable of reacting to produce a normal lens, and by means of such experiments it has again been shown that the chemical content of a dead organizer can act as well as can the cells of a living organizer (Lopashov, 1936). It is also clear from these experiments that at the time of eye formation all epidermal cells are competent to react, but that they lose this competence shortly afterwards.

It is interesting that embryonic differentiation may have negative as well as positive aspects. Holtfreter (1938) has shown that when pieces of anuran epidermis are cultured apart from the embryo they always form a slime-producing sucker, and he interpreted this as indicating

C 2

that when the epidermis is attached to the embryo such differentiation is inhibited except at the normal sites of sucker formation.

From many such fragments of information it is possible to conclude, with Spemann, that the development of a normal embryo depends on elaborate chains, or networks, of tissue inductions occurring in proper sequence in time and space. In amphibian embryos attempts have been made to disentangle the earliest links in the induction pattern, and these have been put into diagrammatic form by Mangold (1961; see Fig. 13). However, an even better conception of the vastness and complexity of the pattern of reactions that lie behind normal organogenesis may be obtained from Coulombre's (1965) analysis of present evidence on the formation of the vertebrate eye. Even within this small organ morphogenesis involves elaborate interactions between a large number

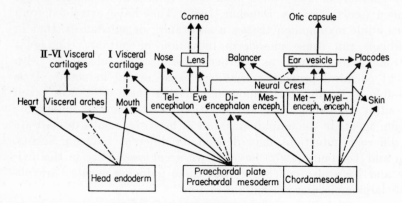

Induction processes in the head

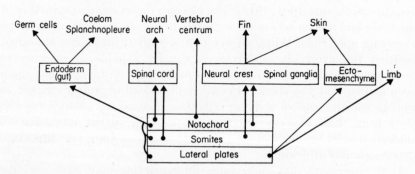

Induction processes in the trunk

FIG. 13. Diagrams showing the induction processes that occur in the head and trunk regions during early amphibian development. (Reproduced with permission from Mangold, 1961.)

of tissues which "assemble in such a manner that their size, shape, orientation, and relative positions meet the precise geometrical tolerances required by the optical function of the organ". Coulombre has summarized the known morphogenetic movements and inductions in a manner which is "necessarily incomplete and oversimplified" and which is shown in Fig. 14.

THE NUCLEUS AND DIFFERENTIATION

The role played in differentiation by inducing agents, whether they are inherited by, produced in, or absorbed into the responding cells, seems to be firmly established, and it is now necessary to consider the role of the nucleus. It is the accepted dogma that most if not all of the information required for tissue differentiation is coded in the DNA of the genes and that the remainder, if any, may be in the cytoplasmic DNA. Ursprung (1965) has used the phrase *"omnis forma e DNA"*, and it is clear that differentiation must depend on the differential interpretation of this DNA-coded information. In other words, since neither genes nor chromosomes are usually lost during organogenesis, the production of a tissue cell from what was originally a totipotent cell must depend on the selective repression of certain gene potentialities and the selective activation of others. Since this happens in response to the presence of inducing agents, it is precisely the type of situation already described as occurring in bacteria and other unicellular organisms in response to the presence of particular effector substances.

Cellular differentiation is usually recognizable because of characteristic visible changes in the cytoplasm, and it is generally agreed that these changes are the result of characteristic syntheses that are initiated in the nucleus. It follows that, since differentiation in the higher animals is at least relatively stable, the characteristic pattern of gene activity must also be relatively stable. Attempts to confirm that gene activity in a differentiated cell is indeed limited to a characteristic set pattern have been mainly made using amphibian embryos. However, in such material, changes in nuclear activity cannot usually be seen and there are a number of other examples, which all relate to the differentiation of germ cells, in which nuclear changes are visible.

Perhaps the best example of visible change is that provided by the developing eggs of the gall midge *Myetiola*, described by Bantock (1961). The early nuclear divisions of the zygote produce a syncytium, and as in all syncytia the mitoses are synchronous. After the fourth successive mitosis, two of the nuclei migrate to the posterior end of the egg and become separated by cell walls. The fifth division within the syncytium then involves only the remaining fourteen nuclei, and at the metaphase

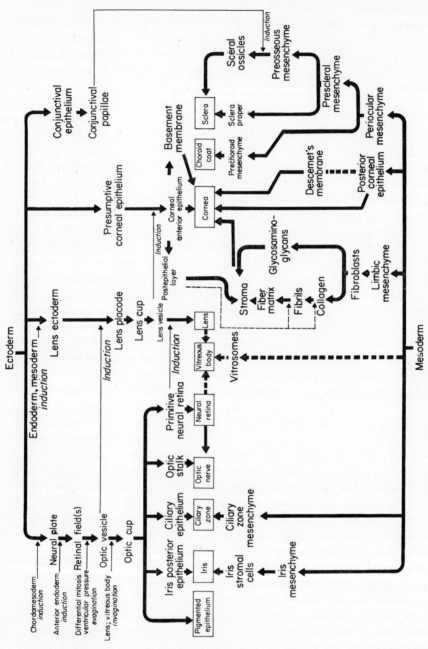

Fig. 14. Diagram showing the major steps in the morphogenesis of the vertebrate eye. (Reproduced with permission from Coulombre, 1965.)

they each lose thirty-two chromosomes, which fail to leave the equatorial plate, become necrotic, and disappear. The resulting twenty-eight nuclei each contain only eight chromosomes, and these direct the whole of the subsequent somatic differentiation (see Fig. 15).

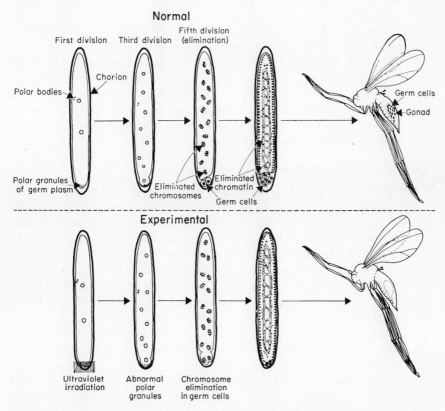

FIG. 15. The process of germ cell formation in the gall midge *Myetiola* in normal and experimental conditions. After ultra-violet radiation the germ cell nuclei are not protected from chromosome elimination and the result is a sterile fly. (Reproduced with permission from Fischberg and Blackler, 1963.)

The two posterior nuclei, besides being isolated within cell walls, are embedded in a distinctive type of cytoplasm that includes many mitochondria and many basophilic granules containing RNA. When these two cells later divide each retains its full complement of forty chromosomes, and the resulting cells give rise to the gametes. It appears that the cell walls together with the distinctive cytoplasm protect the two nuclei from some cytoplasmic influence, which may presumably be regarded as an inducing agent, and which directs the process of chromosome elimination. In support of this it has been found that when the posterior cytoplasm is damaged by ultra-violet radiation it loses its protective properties, so that when the two nuclei arrive they too undergo a mitosis in which they lose thirty-two of their chromosomes. The result is the development of superficially normal but infertile flies.

A closely similar chromatin loss is well known to occur in the presumptive somatic cells in the early embryo of the nematode *Parascaris* (Boveri, 1899; Pasteels, 1948), and again it has been suggested that this is due to the action of some cytoplasmic factor. This factor is evidently missing from, or is counteracted in, the germ cell cytoplasm, which is again distinctively basophilic. Further cases of visible changes in nuclei that are associated with differentiation, especially in plants and insects, are reviewed by Huskins (1947) and Schultz (1952).

EMBRYONIC DIFFERENTIATION IN AMPHIBIANS

It is interesting that in the amphibians the presumptive germ cells also contain a basophilic cytoplasm, although neither they nor the presumptive somatic cells undergo any visible chromosomal change. This basophilic cytoplasm has been described as originating near the cortex at the vegetable pole of the fertilized egg, and in the blastula stage it becomes segregated within certain cells which then "behave differently from those cells without it, as revealed by differences in mitotic rate, cytological appearance and migratory ability" (Fischberg and Blackler, 1963).

Visible differences in the cytoplasm accompanied by visible differences in the nucleus are rare during embryonic development, and the problem of whether relatively stable cytoplasmic differentiation always depends on relatively stable nuclear differentiation has to be attacked by less direct methods. A now classical attempt to do this was made by Briggs and King (1952, 1957, 1959, 1960) using the large eggs and embryos of *Rana pipiens*. The technique involved the extraction of nuclei from embryonic cells and their transplantation into enucleated eggs. In this way it was possible to introduce nuclei in various stages of differentiation into a cytoplasm that is known to be able to support full embryonic development. The quality of the development which then took place under the instruction of the transplanted nucleus was used as a measure of the degree of stable change that the nucleus had previously undergone.

The results of this work appeared clear. If the nuclei were taken from the cells of the blastula or early gastrula they were usually able to direct normal embryonic development. However, if the nuclei were taken from the presumptive endoderm, or chorda-mesoderm, or neural plate cells of the late gastrula, the results were mixed, some nuclei failing to direct development beyond the blastula and others directing essentially normal development. From a range of such experiments Briggs and King (1957, 1960) concluded that the capacity of nuclei to promote normal development falls rapidly in the post-gastrulation period. In particular "the endoderm nuclei undergo stabilized or 'irreversible' changes during differentiation", which appear "to restrict first the ability of the nuclei to promote the formation of ectodermal and ectomesenchymal structures" and "later the capacity to promote gastrulation and chorda-mesoderm formation". Their general conclusion was that differentiation in *Rana* involves increasingly irreversible restrictions on the power of expression of the genome.

These nuclear transplantation experiments have been repeated and extended by Fischberg and Blackler (1961, 1963) and by Gurdon and others (1963, 1964, 1966), using the eggs and embryos of the South African toad *Xenopus laevis*. In general, similar results have been obtained and Gurdon (1964) comments that "there is complete agreement from all nuclear transfer experiments that transplanted nuclei from differentiating cells promote abnormal development more often than do nuclei from early embryonic cells". However, although normal development is not usually obtained with older nuclei, Gurdon questions whether this can be interpreted to mean that the nuclei of differentiated cells have suffered a stable change involving gene repression. The very small cells of fully differentiated tissues are much more difficult to handle than are the blastula cells, and the nuclear changes that have been described and that are evidently stable may have arisen as a result of injury during transplantation. It has, in fact, been noted that some of the stable changes recorded for the transplanted endoderm nuclei of *Rana* are associated with chromosomal abnormalities, which are not seen during normal embryonic development. It thus remains uncertain whether these induced developmental abnormalities provide any information that is relevant to an understanding of nuclear differentiation.

The main evidence against the development of stable nuclear changes during differentiation in *Xenopus* comes from experiments in which the nuclei from fully differentiated intestine cells from feeding tadpoles were transplanted into enucleate eggs (Gurdon, 1964). Although the percentage success was low it proved possible to rear fertile adult male and female *Xenopus* from such nuclei, and Gurdon and Uehlinger (1966) have concluded that "the processes involved in permitting the activity of those genes required for the differentiation of an intestine cell, and in repressing those not so required, cannot therefore entail the irreversible loss or inactivation of genes needed for the differentiation of other cell types".

Thus the development of stable nuclear differentiation along with the obvious cytoplasmic differentiation remains unproved. However these nuclear transplantation experiments do show that the quality of the nuclear activity is dependent on the cytoplasmic environment, and in particular that the cytoplasm of the egg may promote a new phase of totipotency in the genome. This clearly means that genes that were inactive in the specialized intestine cell were activated by transfer to egg cytoplasm. The converse has also been shown to occur, so that genes which were active in the specialized intestine cell were repressed after transfer into the egg cytoplasm. The example given by Gurdon (1963) concerns the genes that support the synthesis of ribosomal RNA. After transfer to egg cytoplasm these genes are immediately repressed, but they are activated later at the appropriate time in embryogenesis.

Thus in the egg cell there exists one pattern of gene activity while in the intestine cell there exists another. The first pattern gives place to the second under the influence of cytoplasmic changes during normal development, and the second as readily gives place to the first when the egg cytoplasm is restored.

The Briggs and King technique has also been used in experiments involving the transfer of the blastula nuclei of one species to the enucleated eggs of another. In this way Moore (1958, 1960) showed that *Rana pipiens* nuclei maintained for a time in *Rana sylvatica* cytoplasm and then returned to enucleated *pipiens* eggs failed to support normal embryonic development. Hennen (1963) confirmed this result and also found that many *pipiens* nuclei showed chromosome abnormalities after their sojourn in *sylvatica* cytoplasm. Gurdon (1962) obtained similar results using the eggs and embryos of *Xenopus laevis* and *Xenopus tropicalis*.

This information, however, may well be irrelevant to an understanding of the normal reactions of nuclei to the cytoplasmic influences to which they are naturally exposed during organogenesis. Reactions to foreign cytoplasm may involve chromosome damage, due perhaps to faulty DNA synthesis (Moore, 1962), and if so they must be regarded as pathological.

EMBRYONIC DIFFERENTIATION IN INVERTEBRATES

The main conclusions that emerge from a study of amphibian embryology evidently apply equally to other types of animals. An introduction to invertebrate embryology generally is given by Brachet

(1960), and typical examples may be chosen from the molluscs and the insects.

The gastropod mollusc *Lymnaea stagnalis* has been particularly thoroughly studied by Raven (1958, 1963), and he too concludes that the earliest steps in differentiation depend on the uneven distribution of cytoplasmic and cortical constituents. Two main types of cytoplasm are clearly visible, and when they are disorganized by centrifugation abnormal embryos result. The cortex, which is only about 100 Å thick, is rich in RNA and sulphydril compounds, and the differences from area to area are more in the form of a mosaic than of a gradient. In particular this mosaic has six characteristic patches, which have been shown to be related to the positions of the six follicle cells that partly surround the primary oocyte while it is in the ovary. This cortical pattern evidently determines the dorsoventral axis, fixes the position and direction of the cleavage spindles, and so controls the segregation of the different types of cytoplasm in the newly-forming cells. The cytoplasmic and cortical peculiarities of these cells then determine the activities of the genes, and so establish the general pattern of the embryo. Then, from about the time of gastrulation, secondary inducing agents begin to be formed.

The situation in insects, reviewed by Agrell (1964), is more difficult to analyse. The egg is relatively small, heavily yolked and enclosed in an impervious shell, and consequently it is not ideal for experimental purposes. However, it is already clear that the antero-posterior and dorso-ventral axes are established in the ovary, and that when the egg is laid there is a distinction between an inner yolky mass, which can be disturbed by centrifugation to cause embryonic abnormalities, and an outer cortex or periplasm. It seems certain that these eggs also follow the general rule of primary induction by substances inherited in the cytoplasm and cortex and secondary induction by substances that are synthesized later in the differentiating groups of cells.

THE STABILITY OF DIFFERENTIATION

Although it is now reasonable to conclude that any particular type of tissue differentiation in a metazoan is the outward and visible sign of a particular pattern of gene activity and inactivity, the question of the normal stability or instability of this pattern remains unanswered. The evidence from the transplanation of *Xenopus* tadpole nuclei into enucleate eggs merely shows that in appropriate experimental conditions a fully differentiated nucleus *can* revert to totipotency; it gives no indication whether such dedifferentiation ever occurs naturally. Indeed, the whole question of the degree of stability of nuclear differentiation remains largely academic unless it can be shown that, at some point in the long process of embryogenesis, dedifferentiation plays some normal and important role or that, after tissue loss in the adult, dedifferentiation contributes to the natural process of repair.

One important conclusion is, however, possible. From the fact that in organisms as widely different as an angiosperm and an amphibian it has proved possible to reopen experimentally those parts of the genome of a differentiated cell that would normally have remained permanently closed, it appears that in many, if not all, differentiated cells the whole genetic message remains intact. Every type of cell that retains an undamaged nucleus may also retain all the information needed to direct the formation of every other type of cell within the same body, although it may remain quite unable to do this. Thus it appears that the suppression of gene action during tissue cell formation does not normally involve any gene mutilation or destruction.

As more comes to be understood, it seems probable that differentiation may be found to show the widest range from the labile states described in micro-organisms, through the semi-stable states which may characterize the more complex unicellular organisms (such as *Acetabularia* and the cellular slime moulds) and the less complex metazoans (such as turbellarians), to the normally fully stable states found, for instance, in mammalian tissues. Indeed, the degree of stability possessed by any differentiated cell may simply be that which long periods of trial and error have shown to be most advantageous for that particular type of cell.

POST-EMBRYONIC DIFFERENTIATION

The sequence of differentiation in any animal can be compared with a system of classification. At first there is the single group of totipotent embryonic cells which may be likened to a phylum, but soon these cells acquire differing characteristics which allow them to be separated into classes. As differentiation proceeds through successive inductions the cells in each class acquire ever more diverse characteristics, which enable them to be separated into orders and ultimately into genera, and even species, of tissues. Thus differentiation proceeds from the general to the particular, and it can be presumed that each step involves a further limitation in the potentiality of the genome. Any differentiating cell passes from a state of genetic totipotency, through a series of states of decreasing genetic pluripotency, to the terminal differentiation by which the cell is typically restricted to a single function.

This progressive narrowing of cell potentiality can be illustrated by the fate of that group of cells which after gastrulation forms the ectoderm. In a mammal these cells differentiate into nervous tissue, neural crest tissue, and epidermis; each of these then differentiates further, the epidermis forming characteristic local areas such as thickened sole of foot or transparent cornea; later still, certain epidermal cells grow down into the dermis to form the eccrine sweat glands or the hair follicles; and finally, from the walls of the hair follicles there grow the apocrine sweat glands and the sebaceous glands. Thus the complex of epidermal regions and epidermal derivatives is established in a characteristic sequence, and the ultimate result may be regarded as a genus epidermis containing many derivative species.

A process of this type is not quickly completed and indeed, in the mouse the last of the hair follicles does not differentiate until a week or two after birth (Gibbs, 1941). Thus differentiation is by no means confined to the embryonic period, which can only be loosely defined as that period when the main organ systems are established and which in most land animals takes place within the egg. Post-embryonic tissue differentiation probably occurs in most, if not all, metazoans, but the impression is gained that the more mature the animal the less frequent it becomes. An exception to this rule seems to be provided by those animals which undergo a dramatic metamorphosis from a larval to an adult form, and which, at least at first sight, give the impression of passing through a second embryonic period during which the adult body is formed. However, this exception may be more apparent than real (see p. 76).

In some animals differentiation even occurs in the adult stage (see p. 82), where it may be seen either as a normal process or during tissue and organ regeneration. It is not yet clear how widespread this is among metazoans generally, and possibly it may be rare.

DIFFERENTIATION IN YOUNG ANIMALS

The genesis of hair follicles in the skin of young mice has already been mentioned. It seems certain that such hair follicles are formed in response to typical inductions, which seem to originate in the dermis. The new follicles do not appear haphazardly over the body surface but are precisely spaced to form a characteristic pattern. After the mice are a week or two old, however, no further follicles are formed and it appears that either the dermis loses its power to induce or the epidermis loses its power to respond. This loss of ability to create new hair follicles seems to be typical of most mammals, but a striking exception is discussed on p. 81.

A remarkably similar case of a terminal induction in insect epidermis has been analysed by Wigglesworth (1954, 1964), who in 1959 summarized his results as follows: "In the growing larva of *Rhodnius* the epidermal cells retain a certain capacity to differentiate and give rise to other organs. These organs are of two types. One is a dermal gland which secretes a thin protective covering on the outside of the cuticle. The other is a tactile sense organ which consists of an innervated hair arising from a socket at the centre of a little dome of smooth cuticle. The essential parts of both organs are formed by four cells which are the daughters of a single epidermal cell. Now the organs of each type are distributed in a regular way over the surface of the cuticle. Two of the sense organs, for example, are rarely found close together. On the other hand, if there is a wide space without a sense organ, a new one is sure to appear there at the next moult when a new cuticle is formed. If the epidermis is killed by applying a hot needle to a certain area of the abdomen, the surrounding cells multiply and grow inwards to repair the injury. The next time the larva molts this new epidermis lays down a cuticle without sense organs but with the normal distribution of dermal glands. At the following moult the sense organs reappear, spaced at the proper distance one from another."

It is presumed that some local chemical messengers, or inducing agents, are responsible for these epidermal inductions, and that some type of chemical interaction between a gland or a sense organ and its surrounding epidermis is responsible for maintaining a proper spacing between these structures. Some similar mechanism may also determine the spacing of the hair follicles in a mammal. What is certain is that in a mammal for some time after birth and in an insect up to the final ecdysis, a mechanism for further tissue induction persists in the skin and that the epidermal cells remain competent to respond.

These examples from a mammal and an insect also suggest that it is only normally the terminal differentiations that occur so late in development; it is only new species that are created within the tissue genera. The indications are that all the major steps in differentiation are normally accomplished much earlier during the embryonic period, and although any undifferentiated cells remaining in a young animal may be pluripotent they are never totipotent. An obvious exception to this rule is, of course, provided by the germ cells.

METAMORPHOSIS AND DIFFERENTIATION IN INSECTS

A further apparent exception is provided by animals that undergo metamorphosis to the adult stage, the classical example being provided by the holometabolous insects (for example, Lepidoptera, Coleoptera,

Hymenoptera and Diptera). In these animals organogenesis in the egg leads only to the formation of the larva, and a later period of organogenesis is needed for the formation of the adult. The interaction of inducing agents and genes in the production of the thorax of the adult *Drosophila* has been described by Lewis (1964). Agrell (1964) has concluded that "in insects no adequate distinction can be made between embryonic and post embryonic development", while Wigglesworth (1965) has suggested that such insects may be regarded as two almost separate organisms that exist sequentially within the single individual.

Present-day insects range from the primitive ametabolous condition (Thysanura and Collembola), in which all the juvenile organs are modified to form the organs of the adult, to the advanced holometabolous condition (Hymenoptera and Diptera), in which there is considerable histolysis of the larval organs and apparent neogenesis of the corresponding adult organs. The hemimetabolous insects (for instance, Orthoptera and Hemiptera) show an intermediate condition (see Wigglesworth, 1954, 1964).

The ancestors of the holometabolous insects appear to have undergone an evolutionary process of larval specialization which was only made possible by a simultaneous evolutionary elaboration of the normal processes of post-embryonic differentiation. It has often been assumed that when, through such post-embryonic differentiation, an insect in the pupal instar acquires such new organ systems as epidermis, muscles, and alimentary canal, the so-called imaginal cells from which these organs originate must be undifferentiated stem cells that have persisted from the early embryonic period (see Henson, 1946). However, considering the neogenesis of muscles, Agrell (1964) has pointed out that it remains doubtful "to what extent imaginal muscles develop *de novo* from imaginal myoblasts or are reorganised from nuclei within the larval muscle cells," and to this it may be added that even myoblasts have progressed a long way along the path of muscle differentiation. Although no critical experiments have so far been carried out, on present evidence it certainly seems that the groups of imaginal stem cells that remain quiescent within the larva are most unlikely to be genetically totipotent, and that at most each group may normally give rise to only a few different but closely related cell types. Thus any differentiation that occurs during the final ecdysis of a hemimetabolous insect or during the pupal stage of a holometabolous insect may involve only one or two terminal steps.

One point is clear, that the production of adult tissues and organs in the insects generally is initiated by a hormonal change. It is the so-called juvenile hormone, produced by the corpora allata (see Gilbert,

1964), which during the juvenile period ensures the persistence of the juvenile organs and prevents the development of the adult organs. Wigglesworth (1965), considering the epidermis, concludes that this hormone acts directly on the cells "presumably by activating the requisite elements in the gene system".

Wigglesworth has considered the juvenile and adult forms of insects as an example of polymorphism, and he has likened it to another common form of polymorphism, sexual dimorphism. Certainly the juvenile–adult dimorphism of the insects resembles the sexual dimorphism of the vertebrates in that they are both hormonally controlled. In an insect it is possible to induce adult characteristics in a nymph or larva by removing the corpus allatum, and also, at least locally, to induce nymphal or larval characteristics in an adult by inserting the juvenile hormone (see Wigglesworth, 1954, 1964). In a vertebrate it is similarly possible to induce female characteristics in a genetic male and male characteristics in a genetic female by appropriate treatment with an oestrogen or an androgen (see Bullough, 1961). Thus the injection of an oestrogen into a male mammal causes the growth of the mammary glands, and the injection of an androgen into a hen causes the growth of the comb. In such cases, however, tissue inductions do not seem to be involved; the mammary glands and the comb both already exist in an unstimulated condition before the hormones are injected. If the parallel is justified, then in a juvenile insect the adult organs may also exist in an unstimulated condition, although it remains to be determined whether, like the unstimulated mammary gland and comb, they are already fully differentiated.

The only obvious remaining difference between the two situations is that juvenile-adult dimorphism in the insect is sequential while sex in the common vertebrates is not. However, sequential sex dimorphism is known in at least one fish, *Monopterus javanensis*, which begins life as a functional female and later transforms into a functional male (Bullough, 1947). There are even some grounds for believing that sequential sex dimorphism was an ancestral condition of the vertebrates.

The problem of the role of differentiation in the metamorphosis of insects remains unsolved (see Wigglesworth, 1965). However, it does appear that even metamorphosis in its most extreme form may involve far less of the embryonic type of differentiation than appears at first sight, and that it is largely hormone-induced. In both vertebrates and insects there appear to be at least two distinct types of hormone-dependent target tissues. The first type, which is almost fully differentiated, appears to retain two separate sets of tissue genes which are alternatively active according to the hormonal environment. In the vertebrates an example is the skin of man which expresses male or female characteristics according to the relative concentrations of the androgens and oestrogens; in the insects, an example is the skin of the hemimetabolous *Rhodnius* which expresses nymphal or adult characteristics according to the presence or absence of the juvenile hormone. Since they are readily reversible such alternative reactions may be regarded as labile forms of differentiation similar to those of the bacteria.

The second type of hormone-dependent tissue, which is evidently fully differentiated, appears to contain only a single set of tissue genes which, however, are only able to express themselves fully in an appropriate hormonal environment. In the vertebrates, an example is the mammary gland which only grows and becomes functional under the influence of female hormones; in the insects, examples are those larval tissue cells of holometabolous insects that express themselves in the presence of the juvenile hormone and that are destroyed at metamorphosis, and the imaginal cells which only express themselves in the absence of this hormone.

UNDIFFERENTIATION AND DEDIFFERENTIATION

It has been suggested that the imaginal cells that are visible in insects from the first instar may all be partly or even fully differentiated. These cells are, of course, commonly described as undifferentiated on the grounds that they show none of the cytoplasmic characteristics of the functional tissue cells to which they will ultimately give rise. However, enough has been said above to indicate that differentiation is fundamentally a nuclear phenomenon. Whatever its appearance may be, a cell must be regarded as fully differentiated if it retains only a single functional potentiality; thus an indistinguishable myoblast is fully differentiated if it retains only the single potentiality to become a functional muscle cell. The fact that, for the moment, the "muscle genes" are inactive is immaterial, and the situation recalls the definition by Foulds (1964) of the facultative genome as consisting of those genes that are inactive but are readily available for use when the need arises. In an insect the facultative genes of the imaginal cells readily become active as soon as the juvenile hormone is withdrawn.

There is even greater confusion over the term dedifferentiation, which is commonly and wrongly used to mean that a cell has lost the tissue characteristics that it previously possessed. This usually happens, for instance, when a cell reverts to mitosis. However, reversion to mitosis does not normally enable a cell to differentiate once more as part of a different tissue. Only if it can be shown that a cell changes its characteristics in this way can it be accepted that there must have been an intervening period of dedifferentiation. The only clear example of major dedifferentiation involving animal cells is provided by the intestine nuclei of *Xenopus* tadpoles, which reacquired a totipotent genome when inserted into the cytoplasm of an enucleate egg (see p. 71).

DIFFERENTIATION AND ASEXUAL BUDDING

If most of organogenesis occurs in the embryo and if only terminal steps in tissue formation are typical of the juvenile, it would not be surprising to discover that differentiation is rare, or even absent, in most adult metazoans. However, a few clear instances are known and many others are believed to occur, although it will require a great deal of precise investigation before it can be decided whether or not these are genuine. Of course, it is already obvious that potentially totipotent cells do persist to form the germ cells, and from what has been said above it is also clear that some hormone-dependent tissues only take their terminal, and reversible, steps in differentiation in the adult animal.

It is also well known that many metazoans multiply by some form of asexual budding, which is particularly well developed, for instance, among the coelenterates, cestodes, annelids and tunicates. Unfortunately, present knowledge of budding in these groups is derived more from detailed descriptions than from critically designed experiments. In all cases there is a variety of possibilities: that the buds are formed from totipotent embryonic cells which have been preserved unaltered in the manner of the germ cells; that they are derived from differentiated tissue cells which undergo partial, or even total, dedifferentiation; or that they are formed from differentiated tissue cells which do not undergo dedifferentiation and which contribute directly to the tissues of the bud. On present evidence it is usually quite impossible to distinguish between these alternatives, particularly when it is realized that it is quite improper to describe cells as undifferentiated or as dedifferentiated merely because they appear featureless.

However, tentative conclusions may be possible in a few selected cases. Thus the posterior buds of such annelids as the freshwater naids may be derived from cells which have remained undifferentiated, since the posterior end of a worm is the growing point from which new segments are formed. Conversely, in many tunicates it is possible that buds are derived from cells which have dedifferentiated. Thus in such genera as *Botryllus* the ectoderm of the adult gives rise to the endoderm of the bud, while in *Clavelina* the mesoderm of the adult gives rise to all three germ layers of the bud (see Berrill, 1950). These are statements based on purely descriptive accounts, and although they could be taken to indicate the persistence of totipotent embryonic cells, they seem rather to indicate the complete dedifferentiation of adult tissue cells. The fact that organs can arise in buds from germ layers other than those from which they are normally derived in the embryo has often caused some surprise, since it implies the "destruction of the sanctity of the three germ layers" (Berrill, 1950). However, this so-called "sanctity" has always been more a matter of convenience than of reality, and it is obviously destroyed in any cell, such as the intestine cell of *Xenopus* inserted in an enucleate egg, that reacquires totipotency.

THE REGENERATION OF ANTLERS

All such examples of budding, however suggestive they may be, provide only inconclusive evidence. For more critical evidence of the potentialities of adult tissues in an advanced metazoan group it is necessary to turn to the illuminating study of antler regeneration by Goss (1964). The facts are as follows.

One group of mammals, the Cervidae, has in the course of evolution acquired the ability to grow new antlers yearly. Each antler is an outgrowth of the frontal bone and during its growth it is sheathed in normal skin which contains large numbers of hair follicles. When the antler is fully grown the skin dies and falls off, and for the rest of the year the antler consists only of naked bone. When the antler is shed, a small bony pedicle remains and this is quickly covered by skin growing in from the periphery. A new antler then grows by the active mitosis of the periosteal and skin cells, and in the rapidly expanding epidermis great numbers of new hair follicles appear.

There has been a long controversy over the ability or inability of adult mammalian epidermis to produce new hair follicles, and although, after extensive wounding, success has sometimes been claimed (see Breedis, 1954; Billingham, 1958), it is clear that hair follicle production is very difficult to induce after about the time of birth. The liberal production of new follicles in the antler skin is therefore surprising, especially as the follicles are all quickly destroyed as the skin dies. In explanation it could be suggested that the skin around the antler base is abnormal in retaining cells in which the terminal differentiations have not yet taken place. However, Goss (1964) has performed the critical experiment of removing this epidermis and of replacing it, by means of a flap graft, with normal ear epidermis. It is well known that transplanted mammalian skin normally retains the characteristics of its original site, and Goss has also shown that ear epidermis, even when extensively wounded, never forms new hair follicles. However when grafted over a growing antler the ear epidermis is able to produce new hair follicles and sebaceous glands on a grand scale.

There is reason to believe that antler formation depends on typical embryonic inductions, which are seasonally activated by hormones, and that the inducing agents are formed in the dermis. Since most attempts at follicle induction in adult epidermis have failed, it appears probable that such epidermis has lost its genetic potentiality for follicle synthesis. If this is true then before such inductions could occur in normal ear epidermis it would be necessary for the cells to reacquire their lost potentiality for follicle formation, which means that part of their closed genome must be reopened. The important conclusion from these experiments is that, with a suitable stimulus, epidermal cells are able to undergo such limited dedifferentiation, and it is also possible that antler regeneration may involve the partial dedifferentiation of other types of tissue cells. This lends support to the suggestion, made above, that it may be possible to reopen the closed genome of any mammalian tissue cell if only the correct experimental procedure can be devised.

DIFFERENTIATION IN ADULT MAMMALS: HAEMOPOIESIS

Although undifferentiated or dedifferentiated "stem cells" may be relatively rare in the adult stages of the higher metazoans, the example of antler growth in deer shows that they can occur to meet certain special problems. One such special problem may be presented by the various groups of dispersed or wandering cells that are found, for instance, in the vertebrates. In the mammals there are three main groups of these cells and each group must be regarded as a distinct tissue. These are the erythrocytes of the blood; the granulocytes (or neutrophils or polymorphonuclear leucocytes) of the blood and other body spaces; and the lymphocytes that are abundant in the blood, the lymph, and the tissue spaces.

These cell systems have been the subjects of a great deal of research, and although it must be emphasized that final proof is lacking, the generally accepted conclusion is that, to meet the high rate of cell loss, each of these tissues is continually recruited from one or two stem cell populations. According to one point of view the same stem cells give rise to all three tissues, and these cells are either closely related to, or even identical with, the small lymphocyte (see especially Yoffey, 1960, 1964). According to another more generally accepted point of view there are two separate stem cell populations, the one giving rise to the erythrocytes and granulocytes and the other to the lymphocytes and plasma cells (see Lajtha et al., 1964; Boggs, 1966). The latest evidence from experiments in which the stem cell populations have been totally destroyed by radiation and in which new stem cells have then been introduced from a donor animal (see Yoffey, 1966) tend to favour the second point of view.

The cell population from which the erythrocytic and granulocytic systems are replenished is known to be concentrated in the bone marrow. If these cells are indeed stem cells then each must take its terminal step in differentiation when it becomes committed to develop into either a mature erythrocyte or a mature granulocyte. This choice is believed to be made according to a chemical stimulus received by the cell. In the case of erythrocyte formation, which is called erythropoiesis, this stimulus is believed to be exerted by a substance called erythropoietin. In the case of granulocyte formation, or granulopoiesis, the responsible substance has not yet been isolated but the existence of granulopoietin has been presumed.

Erythropoietin is thought to be formed mainly in the kidney and it is certainly produced in particularly large quantities whenever the body is short of oxygen, as for instance after extensive haemorrhage or with increased altitude (see Stohlman, 1959; Jacobson and Doyle, 1962). The result is an increased supply of erythrocytes to the blood (see Lajtha and Oliver, 1960). An essentially similar situation develops in relation to the granulocytes when the body suffers a particularly heavy bacterial invasion. Unusually large numbers of granulocytes are quickly released from the bone marrow in which (unlike the erythrocytes) they are stored, and there follows an unusually high rate of granulocyte production from the stem cells (see Craddock, 1960). The erythrocytes in the blood have only a limited life span after which they are destroyed and their constituent molecules re-used. The life span of a granulocyte depends on when it is called upon to function; soon after destroying an invading organism a granulocyte moves through the intestinal mucosa into the intestine lumen, where it is killed and digested (Teir *et al.*, 1963).

If new erythrocytes and granulocytes are indeed produced by terminal inductions, then these cell systems must differ markedly from normal mammalian tissues which replace their lost cells by mitosis (see p. 95). Further, it seems improbable that either erythropoietin or the hypothetical granulopoietin are typical hormones, and at least for the moment they are best regarded as inducing agents. Some support for this theory of induction comes, first, from the fact that the step from stem cell to either pro-erythrocyte or pro-granulocyte does not involve mitosis (although it is followed by mitosis), and second, from the fact that once they have been formed neither a pro-erythrocyte nor a pro-granulocyte appears to be capable of reversing the step it has taken.

The situation in the lymphocyte system is more complex although in essence it may be similar. A mammal contains enormous numbers of wandering small lymphocytes, which have so little cytoplasm that they appear not to be involved in any active syntheses. These cells are constantly produced in great numbers by the active mitosis of large reticular cells in the lymph nodes, and this obviously implies that they must be constantly eliminated in equivalent numbers.

There are two main theories concerning the nature of the small lymphocyte. According to the first, it is a mature and fully differentiated cell of doubtful function and with only a relatively short life span (see Leblond and Sainte-Marie, 1960; Nossal and Mäkalä, 1962). According to the second, the small lymphocyte may "be regarded as a sort of cellular spore form, reduced to the smallest possible size for convenience of transport . . . throughout the body", which only passes "into its active phase when it encounters the appropriate stimulus" (Yoffey, 1966; and see also Gowans and McGregor, 1965; Nossal, 1965). This stimulus evidently depends on a chance encounter with some invading foreign protein or other large molecule, a situation which can be experimentally created, for instance, by the injection of dead bacteria. According to the first theory such a stimulus is felt not by the small lymphocyte but by the reticular cell, which then gives rise by mitosis to a group, or clone, of plasma cells. According to the second theory the clone of plasma cells arises from the small lymphocyte. In either case the function of the newly formed plasma cells is to synthesize the specific antibody that is needed to counteract the invading antigen.

However, the immediate source of the plasma cell is relatively unimportant compared with the question whether or not its production is achieved by a true terminal differentiation. The impression is certainly given that from some generalized cell of the reticular cell small lymphocyte stock there is produced a type of cell that is so specialized that it can synthesize only one type of antibody. Certainly a group, or clone, of cells which has become so limited in its genetic potentiality, and which can never change to produce any other kind of antibody, may be regarded as constituting a single fully differentiated tissue. If this is the manner of production of a plasma cell from a pluripotential stem cell then it is a step which involves a typical terminal differentiation. As Medawar (1963) has said, "the antigen, like the inducer, is an agent that commits a cell to a certain pathway of differentiation— to one pathway among several that it might have taken, and that lie within the genetic capability of the cell".

However, before this point of view can be accepted or rejected much more needs to be discovered of the mechanism of antibody production; something of the complexity and inadequacy of the present evidence has been stressed by Nossal (1965). Furthermore, this theory of induction and differentiation is in direct conflict with the mutation theory of Burnet (1959, 1966). According to this the lymphocyte population already contains a wide variety of fully differentiated cells, each capable of synthesizing only one or two types of antibody, and each only awaiting the stimulus of the appropriate antigen to begin active synthesis.

It is necessary to end by re-emphasizing that the theory of groups of pluripotential stem cells underlying the three main blood and lymph tissue systems is still only hypothetical, and that in the main it is based only on the fact that the so-called stem cells all *look* alike. However, although the critical experiments are still to be devised, the view that stem cells do exist is so widely held by experienced haematologists that it cannot be lightly dismissed.

REGENERATION OF LOST BODY PARTS

The ability to regenerate lost tissues and organs, which is found in some degree in all metazoans, can be conveniently divided into two main, though no doubt intergrading, categories. The first, which may be termed major regeneration, involves the replacement of whole body regions, as in planarians, or of whole limbs, as in amphibians, and it is a process which, at least at first sight, appears to depend on the dedifferentiation and subsequent redifferentiation of adult tissue cells. The second, which may be termed minor regeneration, is seen particularly during wound healing, when there is no evidence of preliminary dedifferentiation and each tissue replaces its own missing parts. Such minor regeneration is discussed in the next chapter (p. 107).

The literature on major regeneration is far too extensive to be dealt with adequately here, but an introduction has been given by Kiortsis and Trampusch (1965). In general it is agreed that the most spectacular powers of regeneration are possessed by animals belonging to the most primitive phyla, and that in the more complex animals the powers of regeneration have been progressively reduced until in the birds and mammals major regeneration is impossible. Unfortunately, as was also the case with asexual budding, most of the available evidence is descriptive and the main problems that are posed have not yet been attacked by sufficiently critical experimentation. The most important of the principles and of the problems that have emerged can be adequately illustrated by reference to only one topic, the regeneration of urodele limbs after amputation (see Goss, 1961, 1965).

When a urodele limb is amputated there is first a period of wound healing, when the epidermis migrates over the cut surface, and then a period of "dedifferentiation which is manifested by the loss of many cellular characteristics by which tissues are normally distinguished from one another" (Goss, 1961). A group of these "dedifferentiated" cells, known as the blastema, then gathers beneath the epidermis at the tip of the stump. Next the blastema cells undergo active mitosis and finally they become "redifferentiated" into new tissues. The main question is whether or not the blastema cells do undergo true nuclear dedifferentiation, or indeed, whether they are really embryonic cells which have been held in reserve in some part of the body. This question cannot be answered but, from a wide knowledge, Goss (1961) has suggested that the blastema cells do originate from the adjacent tissue cells and that the regenerating structures utilize any type of blastema cell that may be available. This clearly implies that the cells must have reacquired many of the genetic potentialities that they lost in embryonic and larval life, and that they may subsequently redifferentiate to form tissues other than those from which they came. However, it must be emphasized that there is still no critical proof that this is indeed true.

The second major problem relates to the ordered way in which the new tissues are formed. As the blastema increases in size the new tissues appear in precisely the correct positions, and clearly there is some morphogenetic mechanism at work to ensure the formation of a properly oriented and integrated limb. This detailed coordination is immediately reminiscent of that seen in the embryo, and it is reasonable to assume that it must be achieved in a similar manner in both cases.

"When a blastema first forms, there is no conclusive evidence that it is in any way determined. Young transplanted blastemas either fail to grow altogether or develop according to their site of transplantation. In older blastemas, however, the cells are determined, with the result that their transplantation elsewhere fails to alter their course of development. These determining factors reside in the tissues of the stump and communicate their influences to the blastema cells" (Goss, 1961).

Certainly without the stump no regeneration is possible, and it seems that the "determining factors" may in fact be typical inducing agents. It is also possible that as the blastema cells multiply and develop they may produce their own inducing agents, and that it is from this stage onwards that the blastema is "determined". This line of argument also requires that the blastema cells must consist either of relict embryonic cells or of tissue cells that have undergone considerable nuclear dedifferentiation.

There is one other major problem that is so far unsolved. It is well known that the regeneration of amphibian limbs and tails becomes impossible if the nerve supply to the blastema is destroyed, and this is also true of other types of regeneration in other phyla (see, for instance, Herlant-Meewis, 1964). Singer (1954, 1958) has suggested that animals such as frogs, which lose their power of limb regeneration as they get older, do so because of an increasing inadequacy of nerve supply. To test this, he amputated the forelimb of an adult frog and increased the nerve supply to the blastema by directing the large sciatic nerve into the stump. In these circumstances the forelimb did regenerate. Singer (1960) has also produced evidence that the "trophic influence" of the nerves is due to the presence of acetycholine, or some related compound, and Goss (1964) has concluded that this nervous influence acts primarily on the blastema epidermis which responds by inducing the accumulation of mesoderm cells beneath it. It has been found that the nerves are only necessary in the earliest stages of limb regeneration, and that once differentiation and morphogenesis have begun a nerve supply is no longer necessary.

However, these conclusions are confused by the results of Yntema (1959). He has shown that if the nerve supply to a urodele limb is destroyed in the embryo a normal but "aneurogenic" limb develops, and that when such a limb is amputated it is fully capable of regeneration. Goss (1964) has concluded that sometime during normal limb development a stage is reached at which the presence of nerves becomes necessary for any subsequent regeneration, but that if no nerves are present the limb "never becomes addicted to them".

Thus the steps in limb regeneration may normally be, first, a nervous induction in the epidermis, second, an epidermal induction in the adjacent dedifferentiated cells, third, a series of inductions emanating from the stump to fix the orientation of the new limb, and last, a series of inductions within the blastema which result in limb morphogenesis.

CONCLUSIONS

This short account of differentiation in the metazoans leads more to problems than to conclusions. Obviously it is here in the metazoans that the mechanisms of differentiation reach their peak of complexity, and the problem posed is probably the biggest and certainly the most important in biology today. However, the one conclusion that does emerge may provide an invaluable clue: there is evidently no essential difference between the labile differentiation of bacteria and the stable differentiation of the metazoans. Wherever it is found differentiation involves the selection of some particular pattern of gene activity that is relevant to the needs of the organism, and in all cases this is evidently achieved as a response to the presence of specific chemical messengers. It is the manner in which this basic mechanism is manipulated that has evolved to meet the needs of different types of organisms.

Monod and Jacob (1961) have summarized the situation. They have pointed out that differentiation in micro-organisms reflects "the relatively unsophisticated regulatory requirements of free-living unicellular organisms, whose only problems are to preserve their intracellular homeostatic state while adapting rapidly to the chemical challenge of changing environments". By contrast the "tissue cells of higher organisms are faced with entirely different problems. Inter-cellular coordination within tissues or between different organs, to ensure the survival and reproduction of the organism, becomes a major factor in selection, while the environment of individual cells is largely stabilized, eliminating to a large degree the requirements for rapid and extensive adaptibility".

GENE CONTROL AT THE MOLECULAR LEVEL

One striking difference between gene regulation in micro-organisms and metazoans is that while in bacteria many of the inducers and re-pressors of gene action are directly related to, or identical with, the metabolites of the pathways which they control, in the metazoans the inducing agents, like the hormones, are evidently specialized molecules that are unrelated to the pathways which they control. However, in neither case is it known how an effector substance acts at the molecular level to modify the activity of the DNA. This problem has been ex-tensively discussed (see especially Bonner and Ts'o, 1964), usually with reference to the possible role of the histones. In short, the argument is that when the acidic DNA molecule is complexed with the basic histone molecule it becomes unavailable for transcription, and that one essen-tial step in gene activation by an effector substance is the separation of the DNA molecule.

The chromosome of any typical eukaryote cell contains centrally a double helix of DNA. The purine and pyrimidine bases that carry the code are at the core of the DNA molecule and they are surrounded by the deoxyribose and phosphate groups, which thus form a sheath. The basic histones (or occasionally protamines) are complexed, at least in part, by salt linkages to these outer phosphate groups, and thus they form a second sheath. The whole structure may then coil and even supercoil.

It is certainly difficult to visualize how the centrally-situated coded message of such a structure could be transcribed unless the DNA molecules were released and even turned inside out, and there is some direct evidence that this may occur. It has been clearly shown by ordinary microscopy that the only parts of the giant chromosomes of insects that synthesize RNA are the swollen regions known as "puffs" (see p. 131). In such regions the DNA molecules are evidently released from their wrappings to uncoil and to loop from the chromosome surface. Similarly electron microscopy has confirmed that repressed DNA forms condensed masses of so-called heterochromatin, while active DNA forms extended fibrils of so-called euchromatin. It is generally believed that such an extension of the DNA molecule could not take place unless its attachment to the histone molecule was broken. The converse of this situation, the inhibition of DNA activity, has also been extensively studied experimentally, and it is now clear that DNA-dependent RNA synthesis can be inhibited by histones (see Huang and Bonner, 1962; Allfrey *et al.*, 1963).

However, although it must be admitted that histone-bound DNA is not active while unbound DNA can be active, it is by no means certain that such binding and unbinding is anything more than an ancillary feature of the normal regulatory mechanism. It is possible that the interactions of DNA and histone are only incidental to the regulation of gene action in a eukaryote cell, and as suggested above, the main role of the histone may be to provide the necessary skeletal support for the very long and very delicate DNA strand.

It is curious how the whole argument concerning the possible regulatory role of the histones has tended to ignore the situation in the prokaryote cell in which the chromosome contains no histone and the cell itself contains very little. In bacteria the hypothesis now in vogue, and repeatedly expressed by Jacob, Monod, and others, is that gene control is effected by repressor molecules which combine specifically but reversibly with their target genes, or operators, and which are believed to be proteins. However, it has never been seriously suggested that these repressors may be histones.

This leads to the further problem of the manner in which a repressor molecule is able to recognize its target gene. A gene is presumably recognizable only by its characteristic sequence of purine and pyrimidine bases, and these are in the most inaccessible position in the core of the chromosome. It is believed that at least the greater part of the DNA-histone binding is ionic and is in fact a phosphate-histone binding. If this is true, and if the histones are the natural repressors of the eukaryote chromosomes, it is certainly difficult to understand how a particular histone could recognize its own target gene. To this may be added the increasing doubt that enough varieties of histones occur within the cell

D

to match specifically the very large number of operator genes that must exist. Considering this difficulty, Butler (1965) has emphasized that a typical cell probably contains a far greater variety of non-histone proteins, which moreover, being non-basic, would probably not become attached to the DNA molecule merely through ionic bonds with the non-specific phosphate groups.

Although repressor molecules are usually considered to be proteins, which alone would fit the allosteric theory of Jacob and Monod (see also Gilbert and Müller-Hill, 1966), the fact that these molecules must in some way accurately recognize a particular sequence of DNA bases has led Frenster (1965) and others to suggest that they may in fact consist of RNA. Such molecules, Frenster suggests, could displace the repressing histones, hybridize with one of the DNA strands, and so free the complementary DNA strand for the synthesis of its specific mRNA. It is not clear, however, how even RNA molecules could recognize their complementary DNA sequences while these remained hidden in the core of the chromosome.

This introduces one of the most difficult problems that stands in the way of any attempt to theorize about regulatory molecules, whether proteins or RNA. As Markert (1965) has said: "Regulatory molecules can scarcely be as varied as the genes themselves, for then the informational potentialities of the genome would be consumed in self-regulation and nothing would be left for coding heterosynthetic functions. Regulatory specificity seems, therefore, to require reliance upon specific arrangements of molecules, rather than upon simply matching specific molecular structure to corresponding genes. We have very little information to guide us in this area, and speculation, being subject to few restraints, is likely to be wrong".

For the present it is only possible to conclude in the most general terms that, as in the bacterial cell, so also in the more complex metazoan cell a gene remains inactive while it, or its operator, is combined with a specific repressor substance, the nature of which is unknown; that the histone of the chromosome acts as a skeletal support for the very long and highly folded DNA molecules, but that it may also play some role in gene repression; that in metazoan cells, as in bacterial cells, the genes are repressed or derepressed in response to the presence of certain chemical messengers, which may be metabolites or specific effectors or inducing agents; and that in a metazoan cell gene activation is accompanied by the breaking of the DNA-histone complex and the uncoiling of the DNA molecule. For a long time the manner in which these reactions are brought about will remain a most important area for research and speculation.

DIFFERENTIATION AND DEDIFFERENTIATION

It has been shown that the activation and inactivation of selected patterns of genes may be immediately reversible as in bacteria, more slowly reversible as during sporulation and encystment in protozoans, and normally irreversible as in mammalian tissue cells. In a mammal the tissue cells are the end-products of a long series of successive steps in cell differentiation, and there is a strong indication that the initial steps in differentiation that took place in the early embryo are ultimately more stable than are the terminal steps that took place in the post-embryonic period. However, even terminal differentiation is normally stable, and it requires abnormal circumstances to induce any sign of instability.

These points are illustrated by the fact that it would appear inconceivable that mammalian epidermal cells could ever become transformed into intestinal epithelial cells, but it is commonly believed that, at least in pathological conditions, certain of the epidermal derivative tissues may be interchangeable. Montagna (1962) in particular has concluded that many epidermal tissue cells are pluripotential: the partial occlusion of the sebaceous gland duct may cause the cells to produce keratin instead of sebum; fragments of hair follicles isolated in the dermis after injury may produce first sebum and later, as pressure increases, keratin; and after epidermal destruction the cells of the sweat gland ducts may migrate outwards and give rise to apparently normal surface epidermis.

However, the most important point about metazoan differentiation is that it is normally so very stable. As Abercrombie (1965) has said: "The results of differentiation, when they can be tested, show in fact a persistence that is not only independent of the environment that originally induced them, but can often be shown not to be diluted out by repeated multiplication. In other words there is a kind of cell or tissue heredity about differentiation, which ensures a lasting change of properties". This would suggest that during metazoan evolution there has been developed a series of repressors which are able to bind particularly strongly with their relevant genes or operators.

However, because the ear epidermal cells can be stimulated by underlying antler growth to produce large numbers of hair follicles, and because the nucleus from a *Xenopus* intestine cell can support normal development in an enucleate egg, it is clear that in appropriate circumstances even differentiated vertebrate cells can recover part or all of their original undifferentiated condition. This removes objections to the suggestion that both asexual budding and limb regeneration may depend on the reopening of previously closed sections of the genome. It also indicates that even firm gene repression need not involve any damage to the genetic code.

THE CELL, THE GENES, AND EVOLUTION

In considering the critically important role of the chromosomal and extrachromosomal genes and the nature of the regulatory mechanisms that control their expression, it is vital always to remember that the basis of life is the cell itself. Not only must the genes be inherited to permit the creation of a new organism but so also must the cell which provides the essential milieu within which the genes can operate.

In essence life consists of the basic metabolic pathways enclosed within the cell. Since all cells from microbes to man depend on essentially the same metabolic pathways, it may also be assumed that within every type of cell there must exist essentially the same group of basic genes, which specify the enzymes on which these pathways depend. These ultra-conservative genes, which probably include most if not all of the extrachromosomal genes, must in some way have avoided all but minor modification since the time when the first cell was established. Evolution has been almost entirely concerned with those other genes which specify and regulate the details of the body that carries and protects the basic mechanisms. Natural selection has led to the increasing competitive efficiency of this body, which has progressed in complexity from a simple prokaryote cell to the integrated multicellular structure of a higher plant or animal. One critical part of this evolutionary process must have involved the regulatory genes that direct tissue differentiation in its proper sequence and pattern.

From time to time in the course of evolution new genes must have been added to the genome and old genes subtracted from it. Many of the new genes must have been newly created, presumably by the lengthening of the DNA thread and the subdivision of the enlarged genes. Others, however, may have been old genes put to new uses.

It is well known that many vestigial organs and processes can be found in both plants and animals, and this must clearly imply the existence of vestigial genes. A vestigial organ or process may be transformed into some new organ or process, as when the chordate endostyle became the vertebrate thyroid, and this may imply a parallel change in the vestigial genes. Alternatively, a vestigial structure or process may ultimately disappear and in this case the vestigial genes may simply lapse into silence. This could occur either by their transformation into nonsense genes or by the development of a particularly firm binding to histone, but in either case they might still remain available for modification and reactivation to serve some new purpose. It is interesting that according to Markert (1965) the DNA content of a single urodele nucleus may be from 7 to 30 times that of a single mammalian nucleus, and yet this can hardly mean that the variety of the amphibian syntheses, or the size of their protein molecules, can be so many times greater than those of a mammal. One conclusion might be that the urodele nucleus contains an unusually large number of old and silent genes. If they are silent merely because of their tight binding to histone it would be fascinating to discover what syntheses they might direct if they could be artificially de-repressed.

CHAPTER 5

Tissue Homeostasis: Mass and Function

Having considered the mechanism of morphogenesis by which a metazoan is created, it is now necessary to consider the mechanism of morphostasis by which a metazoan is maintained in being throughout its post-embryonic life. Although a great deal of research has been devoted to problems of tissue formation, comparatively little attention has been paid to the manner in which tissues are maintained. Perhaps this has been at least partly due to the common attitude that the real problem is the creation of the tissue, which thereafter simply exists. In fact a tissue, once formed, is at all times in a delicate state of balance between the rival forces of cell production, cell function and cell loss. When the situation is examined in detail it is found that both the mass and the functional efficiency are controlled by homeostatic mechanisms, which are also capable of adjusting themselves appropriately to changing circumstances.

It has been concluded that embryonic differentiation depends on the progressive closing of the genome, which is achieved as a series of responses to a sequence of inducing agents. In considering the situation that exists when this process is complete the first important question is what, if any, alternative genetic potentialities still remain in a fully differentiated tissue, and the second is what, if any, is the relation between tissue differentiation and tissue homeostasis. The answers to both these questions vary according to whether the tissue cells are or are not still capable of mitosis, and consequently the problems posed by mitotic and non-mitotic tissues are dealt with separately below. In brief, however, tissue homeostasis may be maintained by the production of new cells by mitosis, by the production of new constituent molecules in existing cells, or by both; the mechanisms controlling cell replacement and molecular replacement appear to be distinct but interrelated.

The problems of tissue homeostasis have been studied more in adult mammals than in any other metazoan group, and consequently the following account deals almost exclusively with the situation in adult mammalian tissues. In these the process has been approached from two main angles. First there is the question of the balanced replacement of tissue cells by mitosis, which is considered below. Second, there is the question of the way in which the tissue cells adjust themselves to the varying work load, and this is considered on p. 117.

MITOTIC TISSUES

In a mammal the mitotic activity of the adult tissues varies widely from the high levels seen in the crypts of Lieberkühn and in growing hair bulbs, through the moderate levels seen in epidermis and in lung alveolar epithelium, to the low levels seen in liver and in kidney (Leblond and Walker, 1956; Bertalanffy and Lau, 1962). There also exist a number of tissues, such as the hypodermis of the skin, in which mitoses are rarely if ever seen but which are capable of rapid cell division after damage, for instance, by wounding.

The group of mitotic cells within a tissue is potentially immortal, but the group of functional cells has only a limited life expectancy which varies widely in different tissues (Bullough, 1965). In general, the ageing and death of tissue cells takes place quickly when the mitotic rate is high and slowly when it is low. Each tissue has its own characteristic rate of ageing, which is evidently genetically controlled (see p. 120).

In all types of tissue the phases of mitosis are closely similar and so also are the phases of cellular ageing. The implication is that in all the cells of the body the same genes direct mitosis and the same genes direct ageing. Thus cellular homeostasis may be dependent on the same mechanism in all cell types, and it is only the point of balance between the rate of cell production and the rate of cell loss that is tissue-specific.

The tissue genes, which of course are also tissue specific, are activated only in the ageing cells, and the intensity of tissue function is determined partly by the number of ageing cells present and partly by their rate of tissue enzyme syntheses, which from time to time may be adjusted to the demand (see p. 123). This situation is summarized in Fig. 16, from

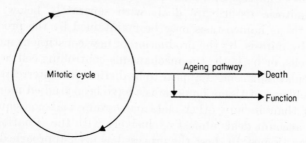

FIG. 16. Diagram of the situation in a typical mammalian tissue. The cells are involved either in the mitotic cycle or in the ageing pathway leading to death. While traversing the ageing pathway the tissue genes are also activated. The rate of tissue function may also vary according to the demand.

which it can be seen that the cells of any mitotic tissue pass at least two critical points on their way from mitosis to death. The first is the point reached after mitosis when the decision is taken whether to prepare for mitosis again or whether to enter the ageing pathway. If the choice is for entry into the ageing pathway then this leads to the second critical point at which the synthesis of the tissue-specific enzymes begins.

Viewed in this way, the cells of any mitotic tissue have two main programmes of protein synthesis, or two main states of differentiation, open to them. They may either synthesize the enzymes for mitosis or they may synthesize the enzymes which determine the composite programme of ageing and function. Although the process of ageing has not previously been emphasized in this way, it has for a long time been well known that differentiation for mitosis and differentiation for tissue function tend to be mutually exclusive (see Bullough, 1965).

CELLULAR HOMEOSTASIS: THE MITOTIC CYCLE

Any cell emerging from mitosis must make the choice whether once more to prepare for mitosis under the control of what may be called, for convenience, the "mitosis operon", or whether instead to prepare for ageing and tissue function under the control of the "ageing operon" and the "tissue operon". Sometimes this choice is taken rapidly as in the lining cells of stomach and intestine, but sometimes it is taken more slowly as when the basal cells of the epidermis remain uncommitted for days or weeks. The period of choice, whether short or long, has been called the dichophase.

The preparations for mitosis involve a series of gene-directed syntheses which are remarkably uniform in all cell types. In any one species it may be safely assumed that the genes comprising the "mitosis operon" are identical in all the cells of the body so that, for instance, mitosis in the intestinal lining is controlled by the same genes as mitosis in the epidermis.

An analysis of the main phases of the mitotic cycle has been made by Bullough (1963, 1965; and see Fig. 17). On leaving the dichophase the

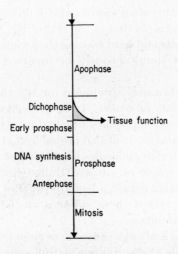

FIG. 17. The phases of the mitotic cycle. From the dichophase the cell may differentiate either for tissue function or for mitosis. (Reproduced with permission from Bullough, 1965.)

cell enters the period of prosphase during which all the syntheses necessary before the cell can enter mitosis are completed. The nature and sequence of these syntheses is still only vaguely understood but it is already clear that prosphase can be divided into at least three main periods: early prosphase, the phase of DNA synthesis (commonly called S), and antephase (commonly called G_2).

The initial syntheses take place in early prosphase, when accelerated RNA and protein production has been demonstrated (Lieberman and Ove, 1962; Hotta and Stern, 1963). It has been suggested that DNA synthesis cannot begin without an adequate concentration of DNA polymerase and that a second enzyme is also needed to "prime" the

D 2

DNA before it can respond to the polymerase (Busch *et al.*, 1963; Mazia, 1963).

Once DNA synthesis does begin it seems that the whole mitotic sequence must pass to completion, although it may sometimes be temporarily blocked in antephase. Certainly the phase of DNA synthesis itself is an all-or-none reaction (Mazia, 1963). This suggests that one of the syntheses of early prosphase may act as a trigger, and also that most, if not all, of the essential gene-directed syntheses must be completed before DNA duplication begins. There is evidence that DNA duplication may pass in a linear manner along the lengths of the chromosomes, and that the rate of DNA synthesis may reach a maximum towards the middle or end of the phase (Brodsky *et al.*, 1964; Kasten and Strasser, 1966). The same studies also indicate that as the rate of DNA synthesis reaches a maximum the rate of RNA synthesis reaches a minimum, and it is reasonable to assume that when DNA is involved in duplication it cannot simultaneously be involved in the synthesis of messenger RNA.

As the cell passes towards antephase (or G_2) the rate of RNA and protein synthesis again increases. It is probable that the cell now manufactures those macromolecules which, during mitosis, aggregate to form the spindle apparatus. It is also probable that antephase is a time when a store of energy-rich molecules is established sufficient to support the cell through the whole of mitosis (Bullough and Laurence, 1964; Bullough, 1965). For this reason adequate supplies of both glucose and oxygen are essential and if these supplies are not available the cell remains arrested in antephase.

Once mitosis begins it always proceeds to completion, and this is true even in the dying cells of dead animals (Bullough, 1950). The changes seen during mitosis are dramatic: the spindle apparatus is completed, the chromosomes condense, the nuclear membrane breaks down, the nucleolus disappears and the endoplasmic reticulum is disrupted. Thus during the time when the chromosomes are moving apart and a new cell wall is being organized, the whole of the machinery for enzyme synthesis seems to be dismantled, and there is general agreement that both RNA and protein production are at a minimum (Konrad, 1963; Prescott and Bender, 1963; Kasten and Strasser, 1966). This, no doubt, is the reason why all preparations for mitosis, including the payment of the energy debt, must be made in advance, and why, once it has begun, a mitosis proceeds to completion in all circumstances short of cell death.

This brief survey of events in prosphase and mitosis is sufficient to indicate: first, that cell division depends on a long and complex series of gene-directed syntheses; second, that these syntheses always take place in a specific sequence; and third, that this sequence is divisible into the four periods of early prosphase with active RNA synthesis, S phase with DNA synthesis, antephase with a second period of active RNA synthesis, and mitosis with little or no gene activity (see Fig. 18). However,

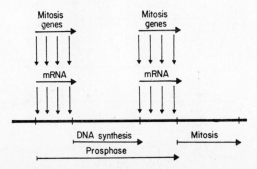

FIG. 18. Diagram illustrating the two periods of high mRNA synthesis during the prosphase, the time of preparation for mitosis. These are also the two periods of maximum sensitivity to chalone inhibition. (Reproduced with permission from Bullough and Laurence, 1966.)

to complete the mitotic cycle a cell must pass through at least one further phase, which is the period of post-mitotic reconstruction that has been called the apophase (Bullough, 1965). Although little is known about it, this is evidently the time when the cell machinery is rebuilt and when the cell grows in mass to recover the size typical of its kind.

From the apophase a cell once more enters the dichophase and once more is faced with the choice between differentiation for mitosis and differentiation for ageing and tissue function.

CELLULAR HOMEOSTASIS: THE AGEING PATHWAY

When a cell enters the ageing pathway it acquires a particular life expectancy that is characteristic of the tissue. When the mitotic rate is high, as in the intestinal lining, the life expectancy of the functional cells is only a few days (Lesher et al., 1961); when the mitotic rate is moderate, as in epidermis, the life expectancy of the functional cells may be a few weeks (Scott and Ekel, 1963); and when the mitotic rate is low, as in liver, the life expectancy of the functional cells may exceed a year (MacDonald, 1961). Thus the flow to death may be rapid, or moderate, or barely perceptible, but it never ceases even in the most stable tissues.

It has recently become clear that the cells of all tissues pass through the same sequence of phases as they move along the ageing pathway to death, and that these phases are under gene control (see Fig. 19). Since ageing as such cannot easily be recognized, these phases have been defined by Bullough and Laurence (1967) in terms of mitosis and tissue function as follows: immature cells (I) are those preparing for tissue

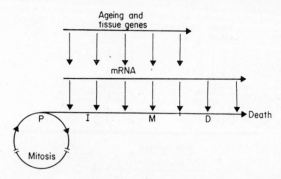

FIG. 19. Diagram illustrating the phases through which all tissue cells pass. *P*, proliferative phase of mitosis; *I*, immature cells preparing for tissue function; *M*, mature cells in which tissue syntheses may be complete; *D*, dying cells. The cells in phases *I* and *M* are controlled by the ageing and tissue genes, while the cells in phase *D* are controlled only by long-lived mRNA. (Reproduced with permission from Bullough and Laurence, 1966.)

function by the synthesis of tissue-specific mRNA and enzymes; mature cells (M) are those in which these syntheses are commonly fully established but which are still capable of reverting to mitosis in an emergency; and dying cells (D) are those in which gene function has ceased and which consequently are moving irreversibly towards death.

In all tissues both the immature and mature cell types are characterized by the fact that the "mitosis operon" remains potentially functional. This is shown, for instance, in wounded epidermis when the cells of the stratum spinosum undergo active mitosis, which they would normally never do (Bullough and Laurence, 1960, 1964); it is shown even more dramatically by the high mitotic activity in the cells of the kidney or the liver during organ regeneration (see Bullough, 1965).

By contrast, cells in the dying phase are characterized by a closed and functionless genome, and consequently they are unable to revert to mitosis in an emergency. The closure of the genome and the continuance of protein synthesis under the control of messenger RNA has been described in lens cells, down feather cells, and reticulocytes (Bishop *et al.*, 1961; Humphreys *et al.*, 1964), in all of which "the nucleus is effectively turned off as . . . the cell progresses towards its terminal state" (Scott and Bell, 1964). The inability of dying cells to revert to mitosis is also demonstrated by the circulating erythrocytes and granulocytes.

FUNCTIONAL HOMEOSTASIS

The impression that the ageing pathway is separable from tissue function is supported partly by the fact that it is present in the same form in all tissues (for non-mitotic tissues, see p. 115) and partly by the fact that tissue function is not always associated with the same stage of the ageing pathway. In tissues with a low mitotic rate, such as liver and kidney, the cells are fully functional in the mature phase, while with a higher mitotic rate, as in the erythrocytic and granulocytic tissues, function is delayed until the cells are in the dying phase. In a few tissues, such as epidermis, the cells are not truly functional until after they are dead, and in certain pathological conditions, such as epidermal psoriasis, death may occur before the tissue syntheses are com-

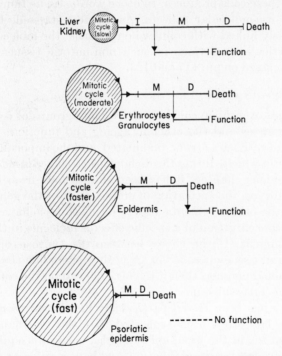

FIG. 20. Diagrams illustrating that the fewer the number of cells involved in mitosis, the slower mitosis is completed, and the longer the cells take to pass along the ageing pathway to death. The greater the number of cells involved in mitosis, the quicker the cells pass to their death, and the later in terms of the ageing pathway the cells become functional. *I*, immature cells preparing for tissue function; *M*, mature cells; *D*, dying cells.

plete (Fig. 20). The special case of function in hormone-dependent tissues is considered on p. 146.

It thus appears that entry into the ageing pathway initiates two separate chain reactions: the first being the sequential gene activation and repression in the "ageing operon" that leads to death, and the second the sequential gene activation in the "tissue operon" that leads to tissue function. The fact that tissue-specific syntheses are under immediate gene control has been shown, for instance, by Davidson *et al.* (1963) in a connective tissue cell line maintained *in vitro*.

Once the "tissue operon" has been activated the genes do not always continue to synthesize mRNA at the same rate. In many tissues, such as kidney and liver (see p. 123), changes in functional demand lead rapidly to parallel changes in the rates of enzyme synthesis (see Goss, 1965). Thus functional homeostasis depends not only on an adequate supply of ageing cells but also on the numbers of enzyme molecules that each of these cells produces. In other words, tissue function has its cellular and its molecular aspects. As already stressed, the cellular aspect is closely linked with cellular homeostasis; the molecular aspect, which is particularly obvious in such non-mitotic tissues as striped muscle, is discussed on pp. 117 and 123.

TISSUE AUTOREGULATION: CHALONES

A typical mitotic tissue consists of two main groups of cells, the one involved in mitosis and the other in ageing and function. Both these programmes of activity are gene-dominated, and the important question is the manner in which, during the dichophase, a choice is made between them. In previous chapters it has been shown how, from bacteria to metazoan embryos, the activation of one segment of the genome to the exclusion of the other segments always seems to be achieved as a response to the concentration of a specific effector molecule in the chromosomal environment. If the choice between the "mitosis operon" and the "ageing and tissue operons" in an adult mammalian tissue is taken in this traditional manner, then it too may depend on the concentration of some intracellular substance.

This theory has been tested particularly in relation to mammalian epidermis. It has often been suggested that the two daughter cells formed by mitosis in the basal epidermal layer are not equivalent but "are unequal in size, form and function, and have a different subsequent development" (Setälä, 1965). One cell is assumed to be predisposed to enter mitosis again, the other to be predisposed to move towards the surface and to synthesize keratin. However, Bullough and Laurence (1964, 1967) have shown that in mouse epidermis the plane of a mitosis is usually such as to ensure that both daughter cells remain in the basal layer. It may be only the increasing pressure that ultimately squeezes cells, apparently at random, into the more distal layers where keratin synthesis begins. Thus in stratified epidermis the dichophase choice between mitosis and keratin synthesis may be made primarily in terms of the position of the cell within the tissue, which may mean that it is taken in terms of the cell environment. Certainly when the cell environment changes after epidermal damage, the cells of the stratum spinosum reverse their decision to synthesize keratin and revert to mitosis (see p. 107).

If environment dictates gene expression in epidermal cells, then the most probable hypothesis is that there exists a concentration gradient of some chemical messenger between the basal and the more superficial layers. The way in which such a system might work in the manner of a negative feedback to suppress mitosis and to confine it to the basal layer is illustrated in Fig. 21; the possible existence in a number of

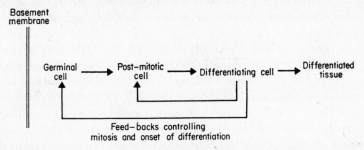

FIG. 21. Diagram illustrating the method of feedback control of mitosis which may operate within the epidermis. (Reproduced with permission from Mercer, 1962.)

tissues of systems of mitotic regulation based on a negative feedback mechanism of this type has often been postulated (see Osgood, 1957, 1959; Glinos, 1960; Iversen, 1961; Mercer, 1962; Bullough, 1962, 1965). Working on this theory, Bullough et al. (1964) have succeeded in extracting from epidermis an antimitotic chemical messenger, which is now called the epidermal chalone and which appears to be a substance of central importance in epidermal homeostasis.

In brief, it has been found that the chalone is tissue-specific but not species-specific. Both in vivo and in vitro, mouse epidermal mitosis can be suppressed by the epidermal chalone extracted from a variety of mammals, including man, and from the skin of the cod fish (Bullough et al., 1967). Thus the epidermal chalone is not even class-specific, and since it must be regarded as a chemical message it follows that the mechanism which emits the message and that which receives and acts upon it must be essentially the same in cod skin as in mammalian skin.

Attempts have been made, especially by Homan and Hondius Boldingh (1965), to isolate and characterize this chalone. The analysis of almost pure samples, obtained from pig skin, suggests that it is a glycoprotein with basic properties and a relatively low molecular weight.

If the epidermal chalone is indeed the effector substance which, according to its concentration within the nucleus, determines the choice between mitosis and keratin synthesis, then a wide range of similar tissue-specific control systems may be expected to exist in the other tissues of the body. It is now known that organ-specific antimitotic substances can be extracted from both liver and kidney (Saetren, 1956; Roels, 1964, 1965), and that a granulocytic chalone is produced by mature granulocytes (Bullough and Rytömaa, 1965; Rytömaa and Kiviniemi, 1967), and indirect evidence also suggests the existence of a hypodermal chalone in mouse skin (Bullough and Laurence, 1960). In the first three cases it has been found that the organ or tissue extract resembles epidermal chalone in being water-soluble, non-dialysable, and heat-labile.

In all these cases it has been suggested that the normal balance between mitosis and function is achieved in terms of the intracellular concentration of the tissue chalone; that this concentration is itself determined by the balance between the rates of chalone synthesis and loss; and that the rate of loss is at least partly related to the difference between the concentration of chalone within the tissue and that in the blood and the body generally. Indeed, the chalone concentration in the blood may be the ultimate factor determining the tissue mass within the total body mass. The tissue may continue to grow until the blood chalone concentration rises to such a point that the tissue chalone concentration is in turn able to rise sufficiently to depress the mitotic rate.

The importance of the tissue mass is also evident *in vitro*. One example is given by Wessells (1964) who, using pancreatic epithelium cultured *in vitro*, has shown that, as growth proceeds, mitosis ceases first in the innermost cells, which then synthesize zymogen granules, and that thereafter it remains peripheral. This is what would be expected if a pancreatic antimitotic chalone was present with its highest concentration centrally and its lowest concentration in the culture medium. It may also be significant that Hauschka and Konigsberg (1966) have shown that the differentiation of striped muscle fibres is promoted by a "culture medium which has been previously exposed, for several days, to the presence of a dense population of cells," and that Cahn and Cahn (1966) have shown that retinal pigment cells lose their pigment when grown as a monolayer and regain it when grown *en masse*. The experimental evidence that cells grown in monolayer always fail to maintain their tissue characteristics, and that a certain minimum three-dimensional mass of cells is needed before these characteristics can reappear, has been reviewed by Davidson (1964).

Since most of the available information comes from mammals, it is particularly interesting that Wigglesworth (1964) has described how tissue mass may control mitosis in insect tissues. He has analysed the mitotic reaction in *Rhodnius* epidermis at moulting time and has shown that it depends neither on nutrition nor on the moulting hormone. The high mitotic activity develops as a "homeostatic response" to the stretching of the epidermis and the consequent wider separation of the nuclei, and "it continues until the cell density normal for the species is restored". He queries whether the nuclei may detect their degree of mutual separation by chemical means, and it is evident that the reaction could be under the control of a similar feedback mechanism to that of mammalian epidermis.

Although more information is needed, the general conclusion must be that cellular and tissue homeostasis in the metazoan body is largely maintained through the actions of a system of tissue-specific chalones. It is possible that there may be many such chalones, but it is also possible that each may control the activities of a range of related tissues. It has already been found that the epidermal chalone will suppress mitosis in oesophageal, corneal and lens epithelia, although hair bulbs, which are epidermal in origin, evidently possess their own control system (see p. 114).

THE ROLE OF THE STRESS HORMONES

It has long been known that mitotic activity in adult mammalian tissues may vary from hour to hour according to a diurnal rhythm. Such rhythms were first described by van Leyden (1916, 1926) in the cat and until recently they have remained unexplained. It is now clear, at least in epidermis, that the mitotic rate is high when an animal rests or sleeps and when in consequence the rate of adrenalin secretion is low, and that the mitotic rate is low when an animal is awake and active and when in consequence the rate of adrenalin secretion is high (Bullough and Laurence, 1961, 1964). Data showing the inverse relation between epidermal mitotic activity and the adrenalin content of the blood are now available for mice, rats, and men (see Bullough, 1965).

Although the diurnal mitotic cycle is best understood in epidermis, it is obvious that similar cycles occur in a wide range of other tissues (see Bullough, 1965). Indeed, it now seems that such cycles exist in most, if not all, tissues that show low to moderate mitotic activity; they are not found in tissues, such as active hair roots, in which the mitotic rate is extremely high. It is also known that stressful situations or injections of adrenalin depress the mitotic rate in tissues with low to moderate mitotic activity, and that they have no effect in tissues, such as active hair roots, in which the mitotic rate is high (see Bullough, 1965).

That adrenalin cannot suppress all forms of mitotic activity indicates that it is not itself an antimitotic substance. In epidermis it is now clear that adrenalin can only prevent mitosis in the presence of the epidermal chalone (Bullough and Laurence, 1964), and that this combined action is further strengthened by the presence of a glucocorticoid hormone (Bullough and Laurence, 1967). It now seems that all types of mammalian diurnal mitotic cycles may be dependent on the joint actions of the tissue chalones and the two stress hormones.

This conclusion explains the absence of diurnal mitotic cycles whenever the mitotic rate is particularly high. A high mitotic rate may be related to a low concentration of the tissue chalone, and without the chalone the stress hormones are powerless to act.

Thus the two stress hormones are built into the tissue homeostatic mechanism, and the choice between the "mitosis operon" and the "ageing and tissue operons" is influenced by the degree of stress felt by the animal.

This raises the question of the effects of the stress hormones on both the rate of ageing and the functional activity of the tissue cells. Concerning the rate of ageing, the only available evidence is from adult mouse skin. It normally takes about 14 days for an epidermal cell emerging from its final mitosis to reach the stratum corneum, where it dies (Weinstein and Scott, 1965). Bullough and Ebling (1952) have shown that, when the mitotic rates of both epidermis and sebaceous glands are sharply reduced by starvation stress, there is no change in the thickness of the epidermis or in the size of the glands, and for this to happen the rate of ageing of both types of cells must have greatly reduced. Thus when chalone action is strengthened by an excess of stress hormones not only is mitosis reduced but the tissue cells are enabled to live and function normally for an unusually long period of time.

This indicates clearly that the rate of ageing is variable and that tissue cells are able to survive in an apparently normal manner for much longer than they usually do. It follows that they must commonly die while they are still capable of further metabolic work, and this in turn suggests that death is not normally due to exhaustion but that the cells are actively destroyed at the appropriate time (see p. 120).

The effects of the stress hormones on tissue function are less obvious, and although an unusually prolonged cell life presumably implies unusually prolonged tissue function, it gives no indication of the intensity of this function. There is some evidence that glucocorticoid hormones may induce an increased rate of enzyme synthesis in liver cells (Caffery *et al.*, 1964; Lang and Sekeris, 1964) and perhaps also in fibroblasts (Berliner, 1964). However, since one function of the liver is to deactivate the steroid hormones, the extra enzyme syntheses may merely be related to the extra metabolic load imposed on the cells (see p. 123).

The question naturally arises why the stress hormones have become involved in tissue homeostasis. One possible answer is that there may be considerable survival value in reduced cell production and increased cell survival in conditions of stress. Probably the commonest form of stress in wild animals is that caused by hunger, when if cell production and cell loss were not reduced the animal might use its reserves more quickly and so die sooner. In other words, since in ideal conditions an animal does not use its cells to the limit, it retains a margin of safety against an emergency.

Viewed in this way the diurnal mitotic cycle, through which this situation was first recognized, may be only an unimportant by-product of the stress reaction.

HOMEOSTASIS AFTER TISSUE DAMAGE

In a mitotic tissue the characteristic response to any form of injury is an increased mitotic rate which leads to the replacement of the lost cells. For a long time it has been widely believed that this response is dependent on the production by the damaged cells, or even by the dead cells, of a stimulating "wound hormone". However, in a series of critical experiments, later confirmed by Finegold (1965), Bullough and Laurence (1960) have shown that it is in fact dependent on the loss of a mitotic inhibitor. Indeed, it was this observation that led to the discovery of the epidermal chalone. After the infliction of a skin wound the mitotic inhibitor, or chalone, is lost only from those epidermal cells lying within about 1 mm of the wound edge, and this loss may be due to reduced synthesis, to dilution within the cell, or to diffusion from the cell. Certainly in epidermal wounds the immediately adjacent cells increase greatly in volume due to rapid water uptake, and this could dilute the intracellular chalone. Also, if this water uptake is a sign of

disturbed permeability in the cell wall there could be considerable chalone loss by diffusion.

In damaged epidermis, mitotic activity reaches a maximum after
about 36 to 48 hours (Bullough and Laurence, 1960). As the mitotic
rate rises the power of the stress hormones to inhibit mitosis is reduced
and the diurnal rhythm fades. This would, of course, be the expected
effect of the disappearance of the chalone, and it too may have a considerable survival value. A wounded animal is usually a stressed
animal and it may also go hungry. This leads to a reduced mitotic
rate and a prolonged cell life in most undamaged tissues, but around the
wound mitotic activity remains high and healing is rapid.

The reactions of cells to tissue damage have also been studied from
another angle by Tsanev (1959, 1962). He inflicted tissue damage by
ligature or, in the case of skin, by pressure, and he found that the
initial response, during the first 3 or 4 hours, is a marked reduction in
the RNA cell content. From this and later work he has developed the
hypothesis that this RNA decrease is the result of the elimination by
ribonucleases of the mRNA which was directing tissue function, and
that the later increase in RNA cell content (which he has also described)
may represent the mRNA needed to initiate mitotic activity (Tsanev,
1964; Tsanev and Markov, 1964; Beltchev and Tsanev, 1966).

In the case of a simple wound the damage is local and consequently
the reaction is local. However, if tissue destruction is widespread
there comes a point at which the local mitotic reaction begins to be
accompanied by a general mitotic reaction affecting the whole of the
undamaged part of the tissue. This response is best understood in the
case of the liver (see Bucher, 1963). A small liver wound leads only to a
local increase in the mitotic rate, but when about 10% of the liver is
destroyed or removed a general increase in mitosis becomes apparent
throughout the whole organ. Beyond 10% the increase in the general
mitotic rate is in direct proportion to the percentage of liver removed
(Bucher, 1963; Bucher and Swaffield, 1964).

The literature on liver regeneration is large and contradictory,
mainly because so much of the experimental work was inadequately
planned (see Weinbren, 1959; Bullough, 1965). In fact the only conclusion which seems to be generally accepted is that the message to commence regeneration must be humoral. This is supported by experiments
both on parabiosis (see Bucher et al., 1951) and on partial liver autografts (Leong et al., 1964). Following the successful joining together
of two rats and the union of their blood systems, it has been,
repeatedly shown that the removal of part of the liver of one rat results
in an increased mitotic rate both in the liver remnant of the operated

rat and in the normal liver of its parabiotic twin. Similarly when one lobe of the liver is grafted into an unusual site, the removal of part of the main liver mass results in an increased mitotic rate both in the liver remnant and in the autograft. Such results are only explicable in terms of blood-borne chemical messengers, but they give no indication of the manner in which these messengers work. Theories have been advanced that such tissue regeneration is due to a temporary shortage of mitotic inhibitors, or chalones (see especially Bullough, 1962, 1965), or to the production of mitotic stimulants, or "wound hormones", by the damaged or functionally overloaded tissue remnants (see especially Paschkis, 1958; Argyris and Trimble, 1964; Goss, 1964), or to both (see especially Needham, 1960).

The available evidence is so confused and contradictory that it is difficult to choose between these three theories. However, Bullough and Laurence (1966) have argued that "organ regeneration is merely an extension on a grander scale of the process of wound healing", which is now believed to be mediated through chalone loss. Certainly wound healing and regeneration share the same essential features: first, they are both sharply tissue-specific; second, they both involve the reversion to mitosis by cells which had previously been synthesizing enzymes for tissue function; and third, in both cases mitotic activity continues only until the tissue loss has been made good. Furthermore, the large literature on mitotic stimulants can be criticized in much the same way as the large literature on "wound hormones". In spite of more than 50 years of effort all such substances remain hypothetical, and until one of them has been extracted and characterized it remains reasonable to question their reality.

With the discovery of chalones it is now possible to explain at least one aspect of the phenomenon of regeneration as follows. When mitotic activity is low, as it normally is in liver and kidney, this is because of the relatively high chalone cell content. The chalone concentration is maintained by the rate of synthesis balanced against the rate of loss, and much of this loss is evidently into the blood (see Fig. 22). With the rate of synthesis constant the rate of loss must be at least partly a function of the steepness of the diffusion gradient from tissue to blood. When a significant part of an organ system is removed the blood chalone concentration will fall, and the result will be an increased loss of chalone from the remaining organ fragment. On the assumption that the rate of chalone synthesis remains constant, the chalone concentration in the tissue cells will then be reduced to a level at which differentiation for tissue function gives place to differentiation for mitosis. The observed relationship between the degree of tissue destruction and the degree of mitotic response is then exactly what would be expected. It is also clear why, when the tissue mass returns to normal, the mitotic activity ceases.

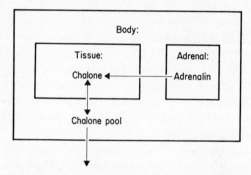

FIG. 22. Diagram indicating the manner in which the chalone concentration within a tissue may be determined by the balance between its rate of synthesis and of loss. The antimitotic action of the chalone is strengthened by adrenalin. (Reproduced with permission from Bullough and Laurence, 1966.)

There are two further points to consider. First, it has often been argued that tissue hypertrophy may be induced by functional overload (see Goss, 1964; Meerson, 1965) and therefore that regeneration, for instance of liver or kidney, may result simply from the extra activity that is imposed on the surviving tissue fragment. This theory is discussed on p. 123. The second point concerns tisues which, like the thyroid, normally grow by mitosis only when under the influence of a specific hormone. As Abercrombie (1957) has stressed, regeneration in such hormone-dependent tissues may be more complex than it is in normal tissues.

THE RATE OF CELL PRODUCTION

So far only the qualitative aspects of chalone action have been considered: when the concentration is low the "mitosis operon" is activated and when it is high the "ageing and tissue operons" are activated. However, it is clear that in any group of cells the chalone concentration may lie somewhere between these two extremes, and in such circumstances "a cell may remain indecisively poised between the two possibilities that are open to it" (Bullough, 1965). A range of intermediate concentrations may perhaps occur naturally within the range of normal adult tissues, but in considering the quantitative aspects of chalone action it is the changes occurring in a single tissue in response to varying degrees of tissue damage or of stress that are most illuminating.

One example is provided by damaged epidermis in which the increased mitotic rate may range from only slight, as seen after simple massage, to very high, as seen alongside a wound (Bullough and Laurence, 1960). This high response, which continues with little or no sign of a diurnal rhythm, shows an increase of more than ten times over the highest normal sleep figure, and it always has two characteristics. First, a cell emerging from one mitosis takes the decision to prepare for a second mitosis much more promptly than does a normal epidermal cell, which means that the dichophase is unusually short, and second, both the prosphase and mitosis are also considerably shortened. Bullough and Laurence (1966, 1967) have concluded that, within limits, any reduction in the epidermal chalone concentration is accompanied by an equivalent reduction in the duration of every phase of the mitotic cycle, which in wounded epidermis may be completed in less than 24 hours.

Indeed, it appears to be a general principle in all types of cells that the lower the chalone concentration the more cells that enter the mitotic cycle and the faster they complete it. In the case of *E. coli*, Rosenberg and Cavalieri (1965) have argued that the rate at which DNA synthesis proceeds is directly related to the quantity of certain enzymes that were synthesized earlier, and possibly the same situation exists in mammalian tissue cells. If this is true then the lower the chalone concentration the more active is the "mitosis operon", the quicker is the onset of synthesis of the essential enzymes, the higher is the final concentration of these enzymes, and therefore the faster is the mitotic cycle. From the evidence of a number of tissues, including hair bulbs (Bullough and Laurence, 1958; Griem, 1966), it appears that in adult mammalian cells the whole mitotic cycle can be completed in about 11 hours, but that this is the limit.

THE RATE OF CELL AGEING

As already mentioned, changes in the mitotic rate are also known to be accompanied by changes in the rate at which cells traverse the ageing pathway. Thus a reduced chalone concentration leads not only to an increased mitotic rate but also to a reduced life expectancy of the functional tissue cells. This has now been shown in several tissues, and from figures given by MacDonald (1961) for rat liver it can be seen that mitotic rate and functional cell life expectancy have an inverse log-log relationship (see Fig. 23; and compare Fig. 31). Conversely, as already mentioned, when the action of the epidermal chalone is strengthened in conditions of stress, the time taken by an epidermal cell to traverse the functional pathway is greatly prolonged (Bullough and Ebling, 1952).

There seems therefore to be a direct quantitative relationship between the chalone concentration and the duration of the ageing pathway. However, the evidence suggests that it is not just the newly formed tissue cells that show an altered rate of ageing. When epidermis is

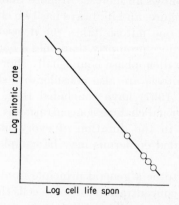

FIG. 23. Graph showing the inverse relationship between the mitotic rate and the functional life span of liver cells. (Reproduced with permission from Bullough and Laurence, 1966.) Compare this with Figure 31.

damaged all the locally affected cells, including those which up to that moment had a relatively slow rate of ageing, immediately acquire a faster rate. Also, when the damage has been repaired, all the new tissue cells, which during the period of high mitotic activity showed a fast rate of ageing, then reacquire the normal slower rate of ageing. Thus the duration of cell life is related not to the mitotic rate at the time when the cells were formed but to the conditions existing at the moment within the tissue. This suggests that both the mitotic rate and the ageing rate may be separately dependent on the same factor, which may be the chalone concentration.

If this is so then, within limits, the degree of reduction in the chalone concentration has a direct quantitative relationship to the degree of activity of both the mitotic and the ageing genes.

This introduces an important safety factor into tissue homeostasis. Because any increase in the mitotic rate is immediately matched by an increase in the rate of ageing, any serious overgrowth of the tissue is effectively prevented. It is for this reason that long continued irritation, which in epidermis, for instance, leads to a continually high mitotic rate, does not also lead to the production of a tumour. The converse is, of course, also true. When the mitotic rate is severely reduced, as it is in epidermis in conditions of stress, it is immediately matched by a reduced rate of ageing so that there is no undue shrinkage or even disappearance of the tissue.

One implication is that after tissue damage the process of regeneration must be completed within the time limit set by the reduced life expectancy of the tissue cells. If this is not achieved then, at the end of this time limit, a new point of balance will be established with the higher mitotic activity exactly matched by the higher rate of cell loss, and from this time onwards no extra tissue can be formed. In fact it has been found, both in healing epidermis and in regenerating liver, that the formation of sufficient new tissue is normally completed well within the time available (Bullough and Laurence, 1967).

As already mentioned, it seems clear from this evidence that cells do not usually die because they are old and exhausted, but that they are deliberately destroyed by some gene-controlled mechanism at a time that is appropriate for the maintenance of cellular homeostasis. It is interesting that even in embryos the cell death rate seems to increase in step with the mitotic rate, and here, too, Saunders *et al.* (1962) have concluded that death must be genetically controlled. Even in early rabbit blastocysts in which the cells are almost completely undifferentiated, so that no tissue function is evident, a raised mitotic rate is matched by a raised cell death rate (Daniel and Olson, 1966). The evidence for cell death in embryos has been reviewed by Glucksmann (1951), and repeated attempts have been made to discover its "developmental utility", for instance in providing space for newly developing tissues and organs. This type of explanation is probably unreal, except in such special situations as the death and absorption of the cells of the tadpole tail during metamorphosis.

This embryological evidence probably merely indicates that the safety mechanism that prevents tumour growth in adult tissues already exists in undifferentiated embryonic cells, and this further endorses the conclusion that the ageing pathway is controlled by a separate mechanism from that which controls tissue function.

UNUSUAL PROBLEMS IN GROWTH CONTROL

Although tissue-specific chalone mechanisms may be of basic importance in the control of normal tissues, it is only to be expected that certain specialized tissues will present particular growth problems that can only be met by the addition of further regulatory mechanisms. This is evidently the case, for instance, in hormone-dependent tissues (see p. 143), and in a different way it is also true of feathers and hairs, which, being mainly dead, are unable to inform the basal mitotic cells when growth is complete. A feather or a hair is formed by the high mitotic activity of a small group of root cells, and this activity ceases abruptly when the structure is fully grown.

The situation is best understood in the case of hair growth, which presents four main phases (Bullough and Laurence, 1958; Montagna, 1962). In the first, the resting phase or telogen, the root cells show no mitosis at all; in the second, called early anagen, mitotic activity leads to the lengthening of the follicle; in the third, called anagen, the root cells undergo continuous high mitotic activity, the duration of which determines the length of the hair; and in the fourth, called catagen, mitotic activity slows and finally ceases and the follicle shrinks to its resting length. In some animals, such as rats and mice, all the follicles in one area act in unison (Durward and Rudall, 1949, 1958), while in others, such as guinea-pigs and men, each follicle follows its own individual rhythm (Chase, 1954).

In the resting phase conditions are similar to those, for instance, in liver or kidney, and the cells are able to respond by mitosis to any form of damage. The impression is that the hair root is held in check by a high chalone content. Conversely, in anagen, the root cells show an extremely high mitotic rate with a mitotic cycle length of only about 12 hr, which approaches the theoretical minimum (Bullough and Laurence, 1958; Cattaneo et al., 1961). The impression is that little or no chalone is present, and in support of this it is found that the cells show no reaction to stress or adrenalin.

It is thus possible that the rhythm of hair growth may depend inversely on a rhythm of chalone production in the root cells (Chase, 1954; Chase and Eaton, 1959; and see Fig. 24), and if this is correct then the stress hormones, although having no effect on either telogen or anagen, may have a marked effect on the transition period of early anagen. If at this time the chalone concentration is falling to permit mitosis to begin, the imposition of stress or the injection of the stress hormones should so strengthen the chalone action that the onset of hair growth is delayed. This effect has in fact been demonstrated during starvation stress (Blumenthal, 1950) and after injections of the pituitary adrenocorticotropic hormone or of cortisone or of adrenalin (Mohn, 1958; Ebling and Johnson, 1964). The converse effect, an acceleration of the initiation of hair growth, has been obtained after hypophysectomy or adrenalectomy.

The rhythmic production of chalone in the hair root cells could depend on some varying factor in the cell environment or it could be an inherent property of the genes. There is evidence that the follicular environment may exert some control over hair cycles (Ebling and Johnson, 1961, 1964), but it seems clear that the main control mechanism lies within the root cells. Bullough (1965) has suggested that this control may depend on "a gene or group of genes which oscillate between activity and inactivity" and which act by modifying the efficiency of the chalone mechanism. Oscillating genetic activity of this kind is already known to control the oestrous cycles of mice (Grüneberg, 1952).

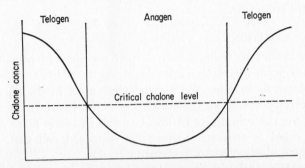

Fig. 24. Diagram illustrating the possible manner of control of the hair growth cycle by the fluctuating chalone concentration in the hair root cells. (Reproduced with permission from Bullough, 1965.)

NON-MITOTIC TISSUES

It has been shown that the balance between cell gain, cell function, and cell loss in any normal tissue is determined by a homeostatic mechanism which selects between two alternative gene-controlled programmes of syntheses. In a non-mitotic tissue, however, no such balance exists. New cells are never created, even in an emergency, and although the tissue may increase in size any such growth is due solely to an increase in cell size. In an adult mammal the best known non-mitotic tissues are the striped and cardiac muscles and the neurones. It is particularly well established in the case of the neurones that cell loss can never be made good, although if an axon is cut the distal region can be regenerated by growth from the proximal region.

In striped muscle the situation is more confused. The review of Betz *et al.* (1966), which describes muscle regeneration with mitosis, fails to distinguish between embryonic and early juvenile muscles which would be expected to act in this way, muscles from such amphibians as newts and axolotls which retain the capacity to regenerate whole limbs, and the adult muscles, for instance, of birds and mammals. In this last group the regeneration of skeletal muscles cut transversely is certainly possible (Lash *et al.*, 1957; Pietsch, 1961), but it is widely agreed that no mitosis occurs in the nuclei of the damaged muscle strands. As the muscle

strands regenerate by growth across the gap, it has been suggested that
the nuclei appearing within them may either have moved in from the
more distal regions of the strands or have been derived from certain
mitotically active mononucleate cells that appear in the wound coag-
ulum. However, none of this evidence contradicts the conclusion that
in the nuclei of the established striped muscle fibres of mammals
mitosis does not occur.

THE BASIC SITUATION

In such tissues as mammalian striped muscle and nervous system
the cells are created by active mitosis during foetal life and this activity
may continue, as in mice, for a short time after birth. Thereafter all
mitotic activity ceases and the stock of cells that has been built up
must last for the remainder of life. Growth in tissue mass during the
juvenile period is achieved merely by increased cell size, a process
which is particularly obvious in striped muscles.

Clearly this situation does not differ in essence from that seen in the
mitotic tissues, in which cells leave mitosis to pass through a mature
phase, when reversion to mitosis is possible, to a dying phase when
it is not. The non-mitotic tissues are incapable of mitosis for one of two
reasons. Either the "mitosis operon" is firmly closed, leaving only the
"ageing and tissue operons" functional, or all these operons are firmly
closed, leaving tissue syntheses to be continued under the control of
long-lived messenger RNA.

Evidence on this point has been provided by Yaffe and Feldman
(1964), who studied gene-directed and RNA-directed syntheses in rat
skeletal muscles *in vitro*. They cultivated mononuclear cells derived
from embryo rat thigh muscles and found that, as *in vivo*, these cells
continued active mitosis for some time. Then, also as *in vivo*, mitosis
ceased and the cells fused together to form long multinucleate muscle
fibres, which developed typical cross-striations and commenced spontaneous contractions.

The synthesis of RNA was then inhibited with actinomycin D both in
the younger mononuclear cells and in the multinuclear muscle strands.
The effects were clear cut. The mononuclear cells all died within a day
or two, but the multinuclear muscle strands remained generally unaffected. Similar results were then obtained with cardiac muscle cells,
the implication being that the younger cells are still dependent on gene-
directed syntheses while the older cells have lost this dependence.

Using tritiated leucine, experiments were then carried out to show
that the older cells which survived actinomycin treatment were still
active in protein synthesis. The final step was to treat such cells with

puromycin, which is known to prevent RNA-directed enzyme syntheses, and the result was that the striped muscle fibres ceased to contract, ceased to synthesize protein, and at the higher doses died in about a day. This suggests that these older cells operate under the sole control of mRNA, and Yaffe and Feldman have concluded that "the functional and morphological properties of differentiated muscle cells are determined by the synthesis of proteins that are formed by a relatively stable RNA".

After mitosis has stopped the passage of the functional cells along the ageing pathway is extremely slow, but the fact that both neurones and striped muscle cells are indeed continually moving towards death has been stressed by Curtis (1963, 1966). He has quoted the estimate "that the human brain loses, without replacement, about 10 000 brain cells every day", and Brody (1955) has also calculated that a man loses about 20% of his neurones by his seventh decade. Whether or not these figures are accurate it is certainly well established that all types of ganglia lose their nerve cells progressively throughout life, and the obvious reduction in size of the skeletal muscles with advancing age is evidently a similar phenomenon.

TISSUE HOMEOSTASIS: MOLECULAR ASPECTS

Although in non-mitotic tissues there is no creation of new cells, there is certainly a constant creation of new macromolecules. In a neurone, molecular synthesis may be so rapid that all the cytoplasm around the nucleus is replaced daily, the older cytoplasm being pushed along the axon at a speed of about 1 mm per day to be destroyed ultimately at the distal end. It appears that there is a similar high rate of mitochondria production and that these structures also pass along the axon to be destroyed distally.

It is also clear that the speed of synthesis of new macromolecules is not always constant, and this is especially obvious in striped and cardiac muscles in which an increased work load induces increased synthesis leading to increased tissue mass. Meerson (1965), for example, has exposed rabbits for 5 hr daily to simulated high-altitude hypoxia (equivalent to 7 500 m above sea level). This caused hyperfunction and hypertrophy of the myocardium so that after 33 days the heart weight had increased by 43%. The experiment was then stopped and it was found that the heart weight returned to normal after about 30 days. Meerson points out that at the beginning of the experiment the hyperfunction was performed by a heart of normal size, which must indicate an increased energy output per unit mass of organ. The resulting hypertrophy then led to a lower energy output per unit mass of organ until the normal ratio was once more reached, and at this point the rate of protein synthesis also returned to normal and growth ceased. When the rabbits were exposed to normal conditions once more this process was reversed with a reduction below normal in the work performed per unit mass of tissue, and a reduction in organ size until the normal balance was again achieved.

Meerson suggests that the mechanism behind these responses may depend on the acceleration or deceleration of DNA-dependent RNA synthesis leading to changes in the rate of protein synthesis, and that "the metabolites associated with cell function may serve as effectors eliminating the repression of regulatory genes". This, of course, clashes with the evidence of Yaffe and Feldman (1964) that in functional cardiac muscles the genes may be inactive, but the reaction could start at the RNA level and it could even be dependent simply on the release of end-product repression.

The conclusion is that non-mitotic tissues contain a homeostatic mechanism which controls the rate of function of the tissue cells in terms of the needs of the moment, and it is clear that at the normal point of balance a tissue possesses a considerable reserve capacity for enzyme synthesis. This is also known to be the case in many mitotic tissues and organs; one kidney, for instance, is capable in an emergency of producing as much urine as is normally produced by two kidneys (Goss, 1965). It follows that tissue homeostasis has two aspects, one of which may be called cellular homeostasis and the other molecular homeostasis, and of these only molecular homeostasis operates in non-mitotic tissues.

CONCLUSIONS

Of the two main homeostatic mechanisms that have been recognized in adult mammalian tissues, the first, which involves a choice between alternative gene-directed programmes of enzyme synthesis, clearly depends on changes in cell differentiation, and there is good evidence that the choice is made in the traditional manner according to the concentration of at least one effector substance, or chalone. The second mechanism does not seem to involve any changes in cell differentiation since it depends merely on alterations in the intensity of either gene or mRNA activity. These two mechanisms, the versatility they confer on the tissues, and their possible interconnections are discussed below.

MITOSIS, AGEING, AND FUNCTION

It seems probable that the basic features of the mechanisms of mitotic and functional homeostasis must have been established during the evolution of the first tissues in the earliest metazoans, and if this is so then mitotic and functional homeostasis is likely to be achieved in fundamentally the same way in all the tissues of all the present-day metazoans. It is also probable that in their origins the mitotic and functional homeostatic mechanisms would have consisted of little more than modifications of already existing mechanisms in the ancestral protozoans. In fact the main innovations may have been, first, the more stable closure of that part of the genome that was not needed by the tissue; second, the production of tissue-specific effector substances to control mitosis in place of the non-specific effector that must have controlled mitosis in the protozoan ancestor; and third, the elaboration of the "ageing operon" and its linkage with the tissue-specific genes.

There are such obvious advantages in tissue stability that the closure of the unwanted parts of the genome has become progressively firmer in the course of subsequent evolution. The ultimate in stability is reached in any tissue in which all the genes are finally silenced and in which all control is exercised at the RNA level. Such a tissue loses the power of cell replacement and retains only the power of molecular replacement.

In many types of tissue, however, there are obvious advantages in the retention of sufficient gene versatility to permit at least the simple choice between differentiation for mitosis and differentiation for tissue function. The cells of the epidermis are constantly worn away, the epithelium of the stomach and duodenum is constantly exposed to digestive enzymes, while the cells of the liver and the kidneys must constantly deal with toxic substances. Cell replacement rather than molecular replacement is essential in such cases, and the choice between mitosis and function is made in the usual way in response to a chemical messenger, or chalone.

A chalone may be defined as an internal secretion produced by a tissue for the purpose of controlling by inhibition the mitotic activity of the cells of that same tissue. This action has three consequences: first, the cells prevented from undergoing mitosis tend instead to enter the ageing pathway; second, once on this pathway the high chalone concentration inhibits the ageing genes so that passage along the pathway is slow; and third, once on this pathway the tissue genes are activated by what may be a simple trigger mechanism, since their rate of activity is not known to be affected by the chalone concentration (see Fig. 16).

In any one tissue the appropriate point of balance between the mitotic rate, the functional rate, and the ageing rate of the cells has evidently been established by selection in the course of evolution, and this presumably means that it is genetically controlled. The simplest explanation may be that the point of balance is achieved in terms of the intracellular chalone concentration, which itself depends on the ratio between the rate of chalone synthesis and the rate of chalone loss by degradation, diffusion, and even excretion. This is certainly an oversimplification, but this ratio is nevertheless likely to be of critical importance. The range of situations that may arise in tissues which contain different concentrations of their chalone is illustrated in Fig. 20. These diagrams also emphasize that, although ageing and function always go together, they depend on separate mechanisms.

Very little is known about the ageing mechanism, which is consequently particularly interesting. An attempt was made by Bullough and Laurence (1966) to define the course of this pathway in terms of tissue function, but since it is now clear that function may occur early or late this definition is seen to be crude and unreliable. There are only two obvious critical events in ageing: first, the closure of the genome as the cell enters the dying phase, and second, death itself. The closure of the genome presumably depends on the tightening of the binding of all the previously active genes to the skeletal histone.

Since at the cell level ageing and function are separate processes, the tissue cells may, and commonly do, die while they are still capable of functioning normally. Ageing and death are delayed in response to the instructions of a chemical messenger which is evidently the tissue chalone; in the absence of this message a cell commits suicide. It seems possible that death itself may be due to the breakdown of the lysosomes, which de Duve (1963) originally called "suicide bags", although it is still not clear whether a cell dies because the lysosome walls break or whether these walls break because the cell dies (de Duve, 1963; Brandes *et al.*, 1965).

In normal mitotic tissues the point of balance between mitosis and ageing-with-function not only allows for an adequate replacement of cells but also enables the cells to react appropriately to the two major hazards to which an animal is naturally exposed, tissue damage and starvation. In wounded tissues healing occurs because the loss of chalone tips the balance towards a higher mitotic rate and a shorter cell life; during starvation the metabolic demands are reduced because the strengthening of the chalone by the stress hormones tips the balance towards a lower mitotic rate and a longer cell life. The changes in the duration of cell life are safety factors which prevent the raised mitotic rate leading to tumour production and the lowered mitotic rate leading to the undue shrinkage of the tissue.

Such reactions to damage and stress are, of course, not possible in those few tissues which possess a naturally high mitotic rate. The most obvious example is the lining epithelium of the duodenum which is supported by populations of mitotic cells in the duodenal crypts. Regarding the crypts, Lamerton (1966) has suggested that the proliferative zone may be "constantly working at full speed and not relying on . . . a feedback control from the mature population" of surface epithelial cells. After wounding these cells cannot divide more quickly than they naturally do; during starvation the stress hormones cannot significantly depress the mitotic rate since there is little chalone with which they can cooperate, although in this situation the high cell loss is more apparent than real since when they are detached into the intestine they are digested and absorbed.

TISSUE HOMEOSTASIS: A MODEL

In all mammalian tissues so far studied the phases of the mitotic cycle are identical and so also are the phases of the ageing pathway. This argues strongly for the existence of a general mechanism controlling cell production and cell loss to which is attached a tissue-specific mechanism controlling function. The connections between these mechanisms are, first, that any cell entering the ageing pathway is activated to commence tissue function, and second, that tissue function leads to the active production of chalone which acts back to reduce the mitotic rate.

Although so little is yet known, it is already possible to discern the probable outlines of these interlinked mechanisms. The basic cell function is mitosis, without which the tissue could not be created. It has already been suggested that since mitosis is an all-or-none reaction it must be initiated by some trigger mechanism and that therefore it is reasonable to regard the genes controlling mitosis as forming a single "mitosis operon" controlled by an operator gene. Following Jacob and Monod it is also reasonable to suggest that this operator may be controlled by a repressor molecule produced by a regulator gene, and that this repressor is inactive in the absence of the tissue chalone.

The obvious difficulty then arising is that the "mitosis operon", together with its operator, must be common to all types of cells, while a chalone is tissue-specific in its action. Of several possible solutions to this problem one of the simplest is shown in Fig. 25. The inactive repressor molecule is produced by a regulator gene, while the chalone is a double molecule, part produced by an effector gene and part by a tissue gene.

In this situation both the repressor molecule and the effector molecule are the same in all types of tissue. It is supposed that the repressor molecule cannot escape through the cell wall, and that the effector molecule is not free but immediately combines with the tissue-specific

E

molecule to form the composite chalone molecule. This composite molecule is tissue-specific and is able to pass through the cell wall, which is also tissue-specific. Since a composite chalone does not affect mitosis in other tissues it is necessary to assume that it is unable to pass through the cell walls of these tissues.

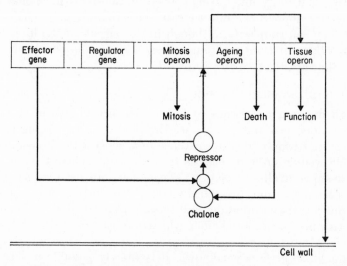

FIG. 25. A possible model of the mechanism controlling mitosis, ageing, and function in a tissue cell. The mitosis and ageing operons, together with their regulator and effector genes, must be common to all types of tissue. Tissue-specificity in mitosis control is obtained by adding a tissue-specific component to the effector (the two together are the chalone) and by the modification of the cell wall. The activation of the ageing operon leads first to tissue function and ultimately to death.

In this system the effector half of the composite chalone could be regarded as being derived from that general effector which controlled mitosis in the protozoan ancestor. Thus tissue-specificity may have been achieved in the earliest metazoans simply by adding a tissue-specific component to control the passage through tissue-specific cell walls.

The repressor is activated when combined with the composite chalone, and this inhibits the "mitosis operon", shifts the cell towards function and simultaneously inhibits the "ageing operon".

The composite chalone thus consists of an active region and a "handle", and this raises the question of the possible chemical similarity of the various tissue chalones. According to the present model they must all contain the same effector and combine with the same repressor; according to the available evidence some, if not all, act better in the presence of the stress hormones; and there is also some inadequate evidence that two or three of them may be small proteins. Thus the evidence of chemical similarity is suggestive but still inconclusive.

FUNCTIONAL HYPERTROPHY

Evidence has been given for the separate existence of a mechanism of cellular homeostasis, by which the cell population of a tissue is determined, and of a mechanism of molecular homeostasis, by which the rate of synthesis of the tissue-specific enzymes is determined. It is now necessary to question if and how these mechanisms are linked. One link is obvious: as the ageing genes are activated, so also are the tissue genes, with the result that the rate of tissue function is partly determined by the number of cells in the ageing pathway. Recently, however, the question has been increasingly asked whether an increase in metabolic demand, which leads to an increase in enzyme synthesis, may also lead to an increased mitotic rate (see Goss, 1964, 1965).

Increased tissue function has two aspects: first, an extra demand on an organ like the kidney arises from an accumulation of extra substrates which must be removed, and second, an extra demand on an organ like the heart arises from an extra need for its products. In other words the metabolic pathways may either be pushed or pulled into higher activity. The extra demand may also arise in two ways: first, a higher functional load may be imposed on a normal organ, as when extra work is demanded from the heart, and second, a higher functional load may be imposed on the surviving remnant of an organ after severe damage, as when part of the liver is removed. The first situation leads to functional hypertrophy and the second to compensatory hypertrophy. Although both must share common features, compensatory hypertrophy is partly dependent on a mitotic response which is due to the reduced cell population.

The critical question is whether functional hypertrophy also involves a mitotic response. The available evidence is contradictory. The negative evidence includes the following: when one ureter is cut there is no extra mitosis in the opposite kidney in spite of the presumed extra excretory burden (Goss and Rankin, 1960); when the salivary glands and pancreas secrete excessively under the influence of pilocarpine there is no rise in mitosis such as follows the partial removal of one of these glands (Alho, 1961; Goss, 1965); and conversely when one of the kidneys is removed from a chick embryo the other kidney shows increased mitosis even though it is non-functional (Ferris, see Weiss, 1955).

The positive evidence includes the following: in the kidney an increased mitotic rate has been described when the functional load is increased by a high protein diet (Argyris, 1966), by changes in the electrolyte balance (Holliday *et al.*, 1961), and in response to acidosis (Lotspeich, 1965); and in the liver an increased mitotic rate has been described when the cells are challenged to detoxicate either methylcholanthrene or phenobarbitol (Argyris, 1966). In both the kidney and the liver the increased rate of enzyme synthesis is gene-dependent, as has been shown by the use of actinomycin D (Goldstein, 1965; Gelboin and Blackburn, 1963) and puromycin (Conney and Gilman, 1963).

Argyris (1966) has concluded that "an increase in liver weight, due in part to mitotic activity, is (always) closely associated with the induction of drug metabolizing enzymes", although he also points out that the "induction of drug metabolizing enzymes appears to precede mitotic activity, and therefore is probably not dependent on it".

However, none of this evidence is conclusive, and in particular it may be suspected that substances like methylcholanthrene and phenobarbitol may damage the cells that are attempting to detoxicate them. If this is so then the mitotic response may be a regenerative response similar to that seen after liver damage induced by carbon tetrachloride.

For the moment the conclusion must be that no firm evidence exists that an increase in tissue function can lead to an increase in mitotic activity. The two mechanisms controlling the rate of cell production and the rate of tissue function are essentially separate and they may also be different in kind.

THE PROBLEM OF AGEING

As an addendum, the consideration of cellular ageing leads to a consideration of the ageing of the whole organism (see Strehler, 1962; Maynard Smith, 1962; Comfort, 1964; Curtis, 1966). There is still no universally accepted theory of the cause or causes of ageing, and the main difficulty arises partly from the apparent differences shown by different types of organism and partly from the conflicting experimental evidence (see Curtis, 1966). The theories that have been advanced fall into two main groups: first, that ageing is due to the random accumulation of damage to the cell machinery until one or other vital tissue or organ can no longer function adequately; and second, that ageing is controlled by genes so that death when it comes is a positive act that is genetically specified. Some of the facets of these two main theories can be summarized as follows:

(i) *Somatic mutation.* The genes of the "tissue operon" may become less efficient with age because of the accumulation of somatic mutations either in the structural or the regulatory genes. These would result in gene malformation leading to reduced synthesis, or to no synthesis, or to the synthesis of abnormal proteins. Such mutations are believed to occur in non-mitotic tissue cells through the random accumulation of damage due to viruses, chemicals, and radiation (see especially Curtis, 1963, 1966).

(ii) *DNA repair mechanism.* It is known that some mechanism exists for the repair of broken or distorted DNA, and that much random damage may be completely eliminated (Dean *et al.*, 1966). The breakdown of the "tissue operon" could therefore be due more to the progressive weakening of this repair mechanism than to the incidence of the damage itself. However, it is also known that some types of gene damage, such as that caused by neutrons, are so severe that they are never repaired. The decreasing efficiency of the "tissue operon" could therefore be due to an accumulation of unmendable gene damage.

(iii) *Auto-immunity.* It has been suggested that the recognition and destruction of cells that have suffered somatic mutation may be a basic function of the immune system. This may provide a second line of defence if the DNA repair mechanism fails to function properly, and such a system may be essential in any animal that lives long enough for somatic mutation to become a serious menace. However, it appears that some somatic mutations may result in the synthesis of proteins that are only slightly abnormal and that in such a case the immune system is only weakly activated. Then, instead of the damaged cell being destroyed, it is attacked in a sublethal manner. These auto-immune reactions become more common with increasing age and they lead, for instance, to rheumatoid arthritis. It is possible that they, too, may contribute to ageing (see Walford, 1962).

(iv) *Gene repression.* It has also been postulated that with increasing age the genes tend to become inactive due to a progressive tightening of their binding with repressor molecules such as histones. Hahn (1966) has suggested that this type of "formation of irreversible bonds is the result of random processes, just as somatic mutations are random". This recalls the final closure of the genome when a tissue cell enters the dying phase (see p. 100), except that in this case the timing of the act is considered to be genetically determined. Hahn has questioned whether such irreversible bonds might perhaps be reopened, and he has pointed out that "no chemical bond, and in particular no enzymatically formed bond, is completely irreversible". Among mammalian tissue cells some reversal is evidently achieved when leucocytes, treated with phyto-haemagglutinin, are able to re-enter mitosis (see Nowell, 1960).

(v) *Programmed genetic ageing.* Most of the theories outlined above are variations on the theme of somatic mutation, and they imply that ageing may be dependent on an entirely random accumulation of gene damage or gene inhibition. However, it is already clear that length of life is genetically influenced and that it may even be genetically determined. Certain strains of mice have a particular expectation of life which is different from that of other strains, and in man longevity runs in families. This clearly suggests that ageing and death occur because the length of life is in some way programmed in the genes, and this is a situation that could be the result of direct evolutionary selection with a certain rate of ageing being advantageous to the species if not to the individual. Thus, in an adult mayfly, fertilization and egg laying are best completed in a day or two, while in a monarch butterfly it is advantageous to migrate south for the winter before returning for egg laying in the following spring.

These various theories are not mutually exclusive and it is in fact possible to amalgamate them. It is also possible to relate them to cell ular differentiation and to cellular ageing. In the first place, it seems probable that genetically programmed senescence and death may be a feature of all the higher animals. In the short-lived adult insects it may be the dominant factor, but in the long-lived mammals its effects may be partly or wholly masked by a random accumulation of genetic damage that leads to death through decreased organ efficiency, increased susceptibility to disease, or the development of such pathological conditions as cancer. In the second place, it seems possible that the programmed senescence and death of the animal may be primarily dependent on the programmed senescence and death of the functional cells of one or more vital tissues. The relation between the ageing of the organism and the ageing of its component cells has not yet been investigated in any detail.

The problem of ageing has been mainly studied in relation to insects such as *Drosophila* and to mammals such as mice and men. Insects pass through a number of immature stages and it is, of course, only in the final, or adult, stage that senescence is seen. Although no exhaustive study seems to have been made, the impression is that insect tissues show no mitoses after the last ecdysis (see Curtis, 1966). Even if this is true only for the majority of tissues, or for the most vital tissues, then their functional cells may be moving at a certain fixed speed along the ageing pathway to death, exactly as occurs in mammalian tissues (see Fig. 19). As the cells enter the dying phase the tissues may depend less on gene-directed syntheses and more on mRNA-directed syntheses, and finally as the cells die the insect must die too. It is therefore possible that in short-lived adult animals it is the genetically specified length of life of the non-mitotic cells of certain vital tissues that determines the duration of the adult stage. In such animals there would not normally be time for the accumulation of any significant amount of somatic mutation.

In a mammal ageing is a more complex phenomenon, and this may be due simply to the long time that is available for somatic mutation to accumulate to a damaging and even fatal degree. However, even if somatic mutation could be entirely avoided the mammal would presumably die, like an insect, according to some genetically determined programme, and there is some evidence that this, too, may be dependent on the speed of ageing of the cells of certain non-mitotic tissues. In particular the cells of the cardiac and skeletal muscle systems, together with those of the nervous system, may be passing slowly but steadily through a protracted dying phase. It has already been mentioned that the human brain may lose about 20% of its neurones by the age of 70 (see Brody, 1955), and the fatal weakening of both muscular and mental capacity is obvious in old age.

Thus in both insects and mammals it may be worth while to investigate more closely the relation between the ageing and death of the cells of vital non-mitotic tissues and the ageing and death of the animal itself.

There remains the question whether the mitotic tissues may also undergo ageing. Certainly their component cells do age and die, but so long as there is adequate replacement the loss does not cause a shrinkage of the tissue and cannot contribute to the death of the animal. However, Hayflick (1965) has considered whether in old age the tissue cells are adequately replaced, or whether in fact the mitotic cells have only a certain limited number of divisions which they can perform. Judging from *in vitro* evidence, he has suggested that human embryonic lung fibroblasts are capable of only about 50 or 60 doublings and that in an adult man only about 20 to 25 doublings remain possible. Thus the ageing of a mitotic tissue could be related to inadequate cell replacement. However, if there is indeed a limit to the mitotic potential of an adult tissue it is probable that this far exceeds the normal life span of the animal. This can be shown in two ways. With long-continued irritation, for instance in epidermis, the raised mitotic rate does not lead to local tissue exhaustion and death; with skin grafts made on to a series of young mice it has been shown that skin can survive in a normal condition for at least three times the life expectancy of the original donor (Krohn, 1962).

It is therefore unlikely that tissue ageing can be related to a loss of mitotic potential, although of course the mitotic rate may be reduced in old age due to the poorer physiological conditions.

However, a mitotic tissue in a long-lived animal is particularly susceptible to the more serious consequences of somatic mutation. In particular, such mutation may so damage the genes that underlie tissue function that the tissue becomes inefficient, or, still more serious, it may so damage the genes that underlie cellular homeostasis that the mitotic rate becomes unnaturally high or the life expectancy in the tissue cells becomes abnormally long. This kind of situation leads to what has been called cellular Darwinian selection, so that the affected cells survive in increasing numbers at the expense of the normal cells and a tumour is the result. This is discussed on p. 155.

CHAPTER 6

Tissue Homeostasis: Hormones

Although the actions of hormones on tissues are dramatic and although more than half a century of research has been devoted to their study, in no case is it known for certain how the hormone molecule acts at the intracellular level. The literature seems to suggest that hormones may perhaps be divisible into two main groups: those that act, perhaps as co-enzymes, on established enzyme systems, and those that act, perhaps in the manner of gene effectors, to induce enzyme neosynthesis (see Tomkins and Maxwell, 1963; Hechter and Halkerston 1965). Hormones that have been thought to act directly on established enzyme systems include insulin and glucagon which control sugar metabolism (Randle, 1964), the gastro-intestinal hormones which cause the release of digestive enzymes (Jorpes and Mutt, 1964), and the catecholamines which have a wide range of actions (Weiner, 1964). However, it is now suspected that insulin may in fact influence carbohydrate metabolism by inducing the gene-directed syntheses of the necessary enzymes (Weber and Convery, 1966), and as Bonner (1965) has remarked "it is becoming increasingly difficult to find a hormone which does not appear to work by derepressing the genetic material".

In many cases this change of emphasis from enzyme reactions to gene reactions is certainly justified, but it also reflects the modern fashion of interest which has itself been swinging in the same direction. In the words of Hechter and Halkerston (1965): "The prevailing conceptual

E 2

fashions in biochemistry and biophysics have dominated the experimental approach to . . . hormone action", and the modern fashion is to produce explanations in terms of gene regulation. Certainly not enough is yet known to allow any safe generalization and the difficulty is increased because the physiological actions of so many hormones are multiple. This may imply that a hormone may have many separate actions within a cell, some perhaps on enzymes and some on genes, but on the other hand it may simply imply that, when the intracellular balance is altered in one particular, the repercussions may extend throughout the whole cell. Multiple actions may thus be the secondary outcome of a single primary action.

In spite of these difficulties, however, one certain conclusion is that a number of hormones are now known to exert at least part of their influence at the gene level, the most important evidence coming almost entirely from studies of insects (Kroeger and Lezzi, 1966) and mammals (see Hechter and Halkerston, 1965).

GENE ACTIVATION BY INSECT HORMONES

In most cases of suspected gene activation or repression the evidence is obtained by indirect methods, and especially from the results of treatment with actinomycin D which is believed to inhibit specifically DNA-directed RNA syntheses. However, in some insect tissues, notably the salivary glands and malpighian tubules of such dipterous insects as *Drosophila* and *Chironomus*, the chromosomes are so unusually thick that regions of gene activity can be sought for visually. The cells of these tissues are particularly large and are engaged in particularly active syntheses. In such cells the normal diploid complement of genes is evidently unable to synthesize sufficient mRNA molecules, and cell function can only be sustained by a polynuclear condition as in striped muscles, a polyploid condition as in some liver cells, or a polytene condition as in these insect cells. To produce the polytene condition the DNA strands replicate repeatedly as the cells increase in size, but the new strands all remain stuck together until some 200 to 4000 of them may lie side by side to form a rope-like chromosome.

The most obvious feature of a giant polytene chromosome is its cross striation or banding (see Fig. 27). Each band is formed by hundreds or thousands of precisely aligned points on the neighbouring DNA strands or chromomeres; one theory is that in each band the chromatin fibres are particularly tightly coiled (see Fig. 26; DuPraw and Rae, 1966). A particular chromosome always shows an identical pattern or sequence of bands and interbands, and it is clear that in their linear arrangement they in some way reflect the linear arrangement of the genes (see Clever, 1964). When one of these giant chromosomes is examined it is commonly found that at a few points it is swollen into what is called a puff

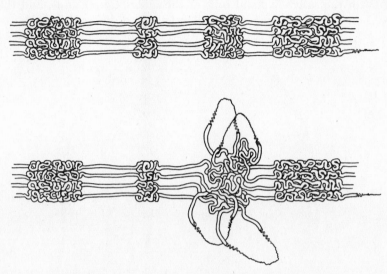

FIG. 26. A folded-fibre model of the band-interband structure of a giant polytene chromosome and (below) the possible method of puffing of one of the bands. (Reproduced with permission from DuPraw and Rae, 1966.)

(see Fig. 27), and it is now believed that each puff represents one or more highly active genes or operons. The slender regions of the chromosome between the puffs are believed to be inactive.

THE SIGNIFICANCE OF CHROMOSOME PUFFING

Each normal DNA thread or chromomere is believed to be coiled and supercoiled and a chromosome puff is believed to be a region of local uncoiling and consequent outlooping (see Fig. 28). By appropriate staining methods, and especially by the use of tritiated uridine, it has been shown that each puff is a centre of active RNA synthesis (Pelling, 1959; Sirlin, 1960), and the fact that this synthesis depends directly on the DNA has been shown by actinomycin D inhibition. It is also known that puffing is not an all-or-none reaction. The size of each puff may vary from small to large and the intensity of the RNA synthesis is in direct proportion to the size. Thus it seems that the more the DNA threads uncoil the higher is their rate of RNA synthesis.

It is an obvious suggestion that what is synthesized may be messenger RNA, and therefore that each puff may produce its own particular type of RNA molecule. Using a microelectrophoretic technique, Edström and Beermann (1962) have determined the base composition of the RNA extracted from four separate puffs on chromosomes I and IV of the salivary gland cells of the midge *Chironomus*, and they have found that this base composition is indeed different in each case.

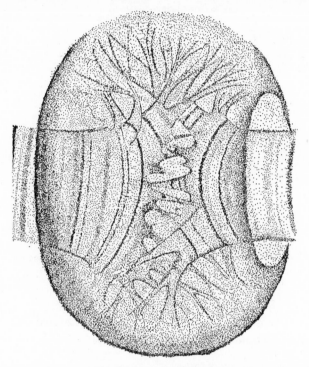

FIG. 27. The structure of a large puff on a giant polytene chromosome. (Reproduced with permission from Beermann and Clever, 1964.)

This supports the view that these are messenger RNA molecules, as does also the evidence from electronmicroscopy that these molecules migrate from the nuclear sap, through pores in the nuclear membrane, to the cytoplasm where they are ultimately degraded.

If the puffs are regions of mRNA production then the pattern of the puffs should be the same in all the functional nuclei of any one tissue, and different in the nuclei of different tissues. It may also be expected that the pattern will change in the nuclei of any one tissue when that tissue changes its functional activity in the course of development. All these expectations have been realized. Within insect larvae "the puffing patterns are identical in all cells having the same function, but vary specifically from tissue to tissue" (Clever, 1964; and see Beermann, 1952; Mechelke, 1953). However, the most impressive evidence comes from the changing pattern of the chromosomal puffs during development. Thus in the salivary gland cells of the larval *Chironomus* there is a regularly repeated sequence of change through the intermoult and moult

FIG. 28. A model showing how a chromosome puff may be formed by the untwisting and outfolding of the DNA threads. In reality a giant chromosome contains thousands of such threads. (Reproduced with permission from Beermann and Clever, 1964.)

periods (see Fig. 29). The beginning of a moult coincides with the appearance of specific puffs, as the moult proceeds other puffs also appear, and at the end of the moult all these puffs regress. In a similar way other puffs wax and wane at particular times in the intermoult period (see Clever, 1964, 1966).

ECDYSONE AND GENE ACTIVATION

Moulting, then, is a process that is characterized by a specific sequence of chromosomal puffs, and it is also a process that is known to be induced by the hormone ecdysone. This is the only insect hormone that has so far been chemically characterized and it has been found to be

134 THE EVOLUTION OF DIFFERENTIATION

a steroid (Karlson, 1963). It is produced by the prothoracic glands which are in turn controlled by a hormone from the brain, and it is known to act directly on the responding cells (Beermann and Clever, 1964). It has been found that a single injection of ecdysone into a *Chironomus* larva is followed within an hour by the development of the same two

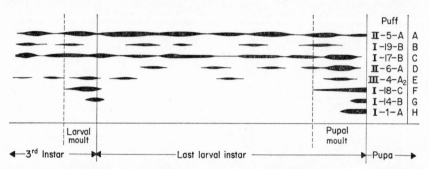

FIG. 29. The time patterns of puffing in certain numbered loci in the salivary gland chromosomes of the midge *Chironomus tentans* towards the end of the larval period. (Reproduced with permission from Clever, 1964.)

chromosome puffs that are characteristic of the earliest stage of the normal moulting period, and the full puffing pattern of the normal moult develops during the following four or five days (Clever and Karlson, 1960; Clever, 1965). It is now clear that the first two puffs develop only when ecdysone is present in the haemolymph and that the size of the puff is proportional to the concentration of the hormone. The concentration required to induce minimal puffing has been established at about 10^{-7} μg per mg larval fresh weight, which represents some 10 to 100 molecules per haploid set of chromatids, while maximum puffing is induced by about 10^{-2} μg per mg larval fresh weight (Clever, 1964).

It is now also clear that the puffs which appear after a few days, and which are characteristic of the later moulting period, do not develop as a direct response to the presence of ecdysone. From a series of experiments Clever (1965) has shown that ecdysone directly activates only one or at most two gene loci, and that it is the products of this activity which by a chain reaction initiate all the subsequent processes of the moult. He therefore concludes that moulting depends not on the activation of a pattern of genes $(A+B+C+etc.)$ but on that of a sequence of genes $(A \rightarrow B \rightarrow C \rightarrow etc.)$ of which only the first is hormonally activated.

The question, however, remains whether or not ecdysone acts directly in the manner of an effector substance to activate the first gene or operon of the sequence. In discussing this Sekeris (1965) has emphasized the speed of the initial response to ecdysone: "the hormone penetrates the cell within 15–30 min, reaching a maximum concentration after 1 hour, stimulation of messenger RNA is seen within 1–2 hours, increase in labelling of microsomal RNA after 3–4 hours and lastly increase of . . . synthesis of enzyme protein after 6–8 hours". This timing may suggest that ecdysone does act directly on the trigger genes. However, as Sekeris also emphasizes, there is no evidence to support this suggestion, and Kroeger (1963) has concluded firmly that insect hormones, including ecdysone, "cannot act directly on the genetic loci of a cell but instead exert their effect on a system in the nuclear sap which in turn controls . . . the genetic loci in the chromosomes". Thus the primary site of ecdysone action remains unknown.

Regarding the chain reaction that follows it is important to note that the details differ according to the stage of development and that the determining factor is another hormone, the so-called juvenile hormone, secreted by the corpora allata. If the concentration of the juvenile hormone is high then ecdysone induces a larval-type moult, but if it is low then ecdysone induces the type of moult that leads first to the pupa and then to the adult. However, ecdysone induces the appearance of the same two initial chromosome puffs irrespective of the concentration of the juvenile hormone, which is therefore believed to act solely by modifying the subsequent chain reaction (see Karlson, 1963).

The next question is whether this chain reaction is mediated at the enzyme level or, as suggested by Clever (1965), at the gene level. In *Calliphora*, the blowfly, part of the chain reaction that develops in the absence of the juvenile hormone involves the production by the larval epidermis of the cuticle-tanning agent N-acetyl dihydroxyphenylethylamine (Karlson and Sekeris, 1962; Sekeris and Lang, 1964; Sekeris, 1965). It is this that causes the formation of the hard brown puparium which protects the pupa, and it is synthesized in the epidermal cells in response to the enzyme dihydroxyphenylalamine (DOPA) decarboxylase. To show that this enzyme is formed *de novo* in response to the synthesis of the relevant messenger RNA, Sekeris and Lang (1964) extracted the RNA from blowfly larvae and tested its ability to direct protein synthesis in an extract of mammalian liver cells. It was found that if the larvae had been treated with ecdysone the enzyme DOPA decarboxylase was formed, but that if no treatment had been given no enzyme was formed.

Thus it is clear that part at least of the chain reaction of the insect moult is dependent on gene activation.

GENE ACTIVATION BY MAMMALIAN HORMONES

The mammalian hormones have a wide range of action and indeed it is probable that at some time during life every mammalian tissue is in some degree hormonally influenced. The chromosomes of mammals, unlike those of certain insect cells, are all so thin that it is impossible to see the regions of genetic activity, although in the lower vertebrates the loops of the lampbrush chromosomes of newt oocytes (see Fig. 10) may be similar in structure and function to the puffs of the insect giant chromosomes.

It is a common feature of all hormones that they are closely involved in the control of functional variations, and thus they play an essential role in homeostatic mechanisms. Although the information is still inadequate it is reasonable to believe that they may operate in a number of different ways. Thus a hormone may act on the physical state of such structures as the cell membranes, which is perhaps how the pituitary hormone vasopressin causes the constriction of blood vessels; another hormone may act directly on an enzyme system in the manner of a cofactor, which is perhaps one feature of the actions of adrenalin, glucagon, and the adrenocorticotropic hormone (Davidson, 1965); and yet another hormone may act by regulating the genetic apparatus. One example of action at the gene level is provided by the mineralocorticoid hormone aldosterone of the adrenal cortex. The primary action of aldosterone is believed to be the regulation of the sodium and potassium ion concentrations (Hechter and Halkerston, 1965). It has now been shown that tritiated aldosterone becomes localized in the nuclei of the responding cells and that the sodium transport effect can be blocked by either actinomycin D or puromycin (Edelman et al., 1963). This has been interpreted to mean that aldosterone acts by stimulating DNA-dependent RNA syntheses, which in turn lead to the syntheses of the necessary enzymes. However, this does not necessarily mean that the hormone acts directly at the gene locus.

In the case of hormones like aldosterone it is clear that their control must be continually exercized by the regulation of existing enzyme systems rather than by the initiation of de novo syntheses. Such hormonal control operates like the governor of a machine to ensure that the speed of the reactions it dominates is appropriate to the needs of the moment.

For present purposes, however, the most interesting hormones are those which induce neosyntheses in tissues which would otherwise remain inactive. Although such a system differs only in degree from the governor system described above, it provides a more dramatic situation which is ideal for experimental analysis. Hormones of this type include the various pituitary and gonadal sex hormones, which determine the various types of sexual cycles, as well as other related hormones such as prolactin, which helps stimulate the growth and function of the mammary glands and the crop glands of pigeons. These are all hormones which typically are produced only during the breeding season and which are related not to the homeostasis of the individual but to that of the species. They are also hormones that cause not only tissue function but also growth by mitosis in their target tissues, and so they may be collectively called mitogenic hormones. The best known are the oestrogens.

OESTROGENIC HORMONES

In terms of their action on the vaginal and uterine epithelia the more obvious actions of the oestrogenic hormones may be summarized as follows: within about 30 minutes there is an increased water uptake and an increased rate of synthesis of both messenger and transfer RNA; within a few hours there is an increased rate of synthesis of ribosomal RNA and a 300% increase in the rate of protein synthesis; after about 24 hr there is a burst of high mitotic activity; and after about 36 hr large numbers of cells begin to die, a process which in the vagina is accompanied by keratinization (Hechter and Halkerston, 1965; Bullough, 1965; Epifanova, 1966). The other vaginal and uterine tissues react more slowly (Bullough, 1946), and indeed, as Hechter and Halkerston have stressed, "given a set of differentiated cell types which are directly responsive to a hormone, it does not necessarily follow that the initiating hormone-receptor reaction in each cell type need be identical, or that the hormone-receptor reaction in different cell types is coupled to the same set of secondary and tertiary reactions".

However, taking the uterus as a whole it has been reported that the early increases in RNA and protein syntheses are blocked by injections of either actinomycin D or puromycin (Gorski and Nicolette, 1963; Hamilton, 1963; Ui and Mueller, 1963), and Davidson (1965) has concluded that "there can be no reasonable doubt that treatment with estrogenic hormones results in activation at the gene level" in the target tissues. He also has pointed out "that a considerable number of genes must be activated in order to account for the many different responses of the cells".

As with ecdysone, this does not, of course, necessarily mean that the hormone itself acts directly at the gene level. From their thorough analysis of the uterine response to oestrogenic stimulation, Hechter and Halkerston (1965) conclude "that the primary effect of the hormone is not on RNA synthesis *per se*" but on a receptor which "in some way stimulates the synthesis of certain specific proteins necessary for the induced synthesis of new RNA". Thus an oestrogen may not directly affect any genes but may react, perhaps allosterically, with "a unique protein—a receptor molecule—which (translates) the message into the language of intracellular informational signals". This agrees with the conclusion of Tomkins and Maxwell (1963) that the stimulus to RNA and protein synthesis could be achieved through hormone-induced changes in the structure of a repressor substance. It is tempting to suggest that this substance might be the tissue-specific chalone which holds the normal balance between mitotic activity, cell function, and the rate of cell death.

Some support for this suggestion comes from Talwar *et al.* (1965), who believe that because "target organs sequester the appropriate hormone from the systemic circulation . . . even when the amount is very low" it is probable that there must exist "specific receptor substances in target cells capable of recognizing a particular hormone". They have in fact demonstrated the existence of a uterine fraction which selectively binds with oestradiol, and they show that when unbound this fraction inhibits RNA polymerase activity but when bound it loses this ability. It is possible that the active substances may be the uterine chalones or at least that they may form an integral part of the chalone mechanism.

The complexity of the uterine reaction to oestrogens has been well described by Hechter and Halkerston (1965), but for present purposes it may simply be noted that the reaction is double. Whether the effect is direct or indirect the hormone activates both the genes for mitosis and the genes for tissue function. It has already been emphasized that mitosis and tissue function are mutually exclusive, but this apparent contradiction is easily resolved. An analysis for instance of the response of the stratified vaginal epithelium (Bullough, 1946) shows that mitosis occurs first in the basal cells, while tissue function (keratin synthesis) occurs later in the distal cells.

An excellent analysis of the mitotic response in uterine epithelium has been given by Epifanova (1966). In ovariectomized control mice only 10% of cells were involved in the mitotic cycle, which had an overall length of 42 hr ($G_1 = 31 \cdot 5$ hr; $S = 8 \cdot 5$ hr; $G_2 = 1$ hr; $M = 1$ hr). After treatment with oestrone 30% of cells were involved in the mitotic cycle, which then had an overall length of only 26 hr ($G_1 = 18 \cdot 5$ hr; $S = 5 \cdot 5$ hr; $G_2 = 1$ hr; $M = 1$ hr). Thus the rise in the mitotic rate is due to more cells completing mitosis more quickly. Epifanova concludes that the main action of oestrone is to accelerate the transition "from

G_1 to S", which is the period here called the dichophase, although to a lesser extent it also accelerates the phase of DNA synthesis.

It may be significant that these are both responses that would be expected in any mitotic tissue when the chalone concentration is reduced. Yet another expected response would be a shortening of the life expectancy of the tissue cells, and this too has been demonstrated in the rat vagina (Peckham and Kiekhofer, 1962). In ovariectomized animals the cells of the vaginal lining may survive for months, but when stimulated by an oestrogen they suddenly acquire a life expectancy of only about 30–45 hr.

Thus the reaction of the female genital ducts to oestrogenic stimulation takes place in a series of stages: first, the increased rate of RNA and protein synthesis which may be related to both the increased metabolic rate and the premitotic syntheses; second, the increased mitotic rate which leads to the increased duct size; and third, the increased flow of cells through the functional pathway to earlier death, which leads to a new point of balance between cell gain and cell loss and so sets a limit to the increase in duct size.

There remains the question of tissue function, and this is best analysed in terms of the vaginal epithelium which produces keratin only in the dying cells. Keratin synthesis cannot be active in an unstimulated epithelium simply because so few of the cells reach the dying phase, but after oestrogen treatment when large numbers of cells pass to an early death, keratin is formed abundantly. This suggests that an unstimulated target tissue is functionally inactive simply because most of the cells remain in the immature or mature phases and because tissue function is only undertaken by cells in the dying phase (see Fig. 30). Once again this leads to the possible conclusion that an oestrogen acts primarily by modifying the chalone mechanism, so altering the balance between cell gain and cell loss.

If it is simply the raised mitotic rate that leads to function through increased cell death, then anything that increases the mitotic rate should have a similar effect. It is in fact well known from studies of vaginal smears that constant vaginal irritation leads to a form of keratinization that is at least superficially similar to that induced by an oestrogen. It is also well known that the effects of wounding are similar and include, both in epidermis and vagina, increased water uptake, increased RNA and protein synthesis, increased mitotic activity beginning after about 24 hr, and increased keratinization. There is also the evidence of Prop (1965) that in subthreshold situations in mammary gland *in vitro* the mitotic stimulus of a hormone and of a wound are additive. For these reasons Bullough (1965) has suggested that at least one of the ways in which a mitogenic hormone, such as oestradiol, may act is by neutralizing the antimitotic action of the tissue chalone, which may then be the "unique key protein" postulated as the receptor molecule

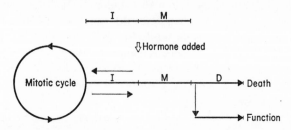

FIG. 30. Diagram illustrating the changes occurring in a target tissue when stimulated by hormone. In the unstimulated condition (above) the non-functional cells show little mitosis or death; when stimulated the cells first revert to mitosis and their progeny then move rapidly along the ageing pathway to death, becoming functional as they reach the dying phase. *I*, immature cells preparing for tissue function; *M*, mature cells; *D*, dying cells.

by Hechter and Halkerston (1965). If this is true then a target tissue is merely one in which the chalone mechanism is especially susceptible to interference by the hormone in question.

ANDROGENIC HORMONES

Unfortunately the manner of action of this second group of mitogenic hormones is not nearly so well known as is that of the oestrogens. However, it is reasonable to assume that the androgens probably act on the male accessory sexual organs in much the same way as do the oestrogens on the female accessory sexual organ. Indeed, the only obvious difference between them lies in their target tissues.

The slender available evidence suggests that in stimulating the growth of such organs as seminal vesicles and prostate the androgens act through the neosynthesis of mRNA and of enzymes (Kassenaar *et al.*, 1962; Kochakian and Harrisson, 1962; Liao and Williams-Ashman, 1962), although in such accessory structures as chick comb it is possible that they may act without such neosynthesis (Talwar *e, al.*, 1965).

Unfortunately in the extensive literature on androgens the term "growth" has usually been used loosely without any proper distinction between hypertrophy and hyperplasia (see Price and Williams-Ashman, 1961; van Oordt, 1963). In fact the only quantitative studies of androgen-induced mitotic activity relate to the non-sexual tissues epidermis (Montagna and Hamilton, 1949; Bullough and van Oordt, 1950) and sebaceous glands (Montagna and Kenyon, 1949; Ebling, 1957).

Particularly interesting are the mitotic counts obtained by Ebling (1957, 1963) from rat sebaceous glands after castration, hypophy-sectomy, and testosterone injection when compared with the sebaceous cell life expectancy. The inverse log-log relationship between the rate of cell gain and of cell loss shown in Fig. 31 appears to be identical with that found in normal tissues when the mitotic rate changes after cell damage (compare Fig. 23).

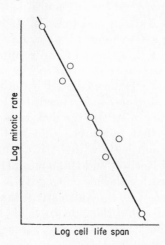

FIG. 31. Graph showing the inverse relationship between the mitotic rate and the life span of sebaceous gland cells in various states of hormonal stimulation. (Reproduced with permission from Bullough and Laurence, 1966.)

Again the important point is made that, although a tissue may en-large due to an increased mitotic rate, the limit to this growth is set by the shorter life expectancy of the cells. In the presence of a mitogenic hormone a target tissue can grow in size only during that period of hours or days before the tissue cells, old as well as new, begin to die in large numbers. From this time onwards a new point of balance is reached between cell gain and cell loss. Also if tissue function is typical of the dying phase of cell life, then an increased cell flow to death is all that is needed to produce an increased rate of tissue function.

GLUCOCORTICOID HORMONES

This is another group of steroid hormones that has been studied for its possible action at the gene level. These hormones have a wide range of actions which affect almost all the tissues of the body. They have been mainly studied in two connections: in the liver it has been found that cortisol, for instance, stimulates RNA and protein synthesis (see Hechter and Halkerston, 1965); in the other tissues of the body such as muscle, spleen (Kit and Barron, 1953), lymphocytes (Blecher and White, 1958; White et al., 1961) and reticulocytes (Koritz and Dorfman, 1956), cortisol inhibits protein synthesis. Thus there is a curious disparity of action between the hepatic and the extrahepatic tissues, which must be considered separately.

In the liver, cortisol evidently induces the neosynthesis of a wide range of enzymes, many of which are concerned with gluconeogenesis from amino-acids (see Karlson and Sekeris, 1966), and the delay in induction ranges from a few hours to a few days, which suggests the existence of a chain reaction. Most of the results have come from *in vivo* experiments and are therefore open to the possible objection that they may be secondary effects mediated by other tissues. However, Goldstein et al. (1962) using an *in vitro* perfusion technique, have shown clearly that cortisol acts directly on the liver to induce the neosynthesis of enzymes, and also that puromycin blocks this response. Similar inhibitions of enzyme synthesis with both actinomycin D and puromycin *in vivo* have led Hechter and Halkerston (1965) to conclude that there is "good evidence for the messenger-RNA participation in neoprotein synthesis". However, they also stress that, as with the oestrogens, there is no certainty that the glucocorticoid hormones act directly at the gene locus rather than indirectly through some intermediate mechanism.

In considering the synthesis-promoting action of a glucocorticoid hormone in the liver it is important to recall that one major function of this organ is to deactivate the steroid hormones just as it also detoxicates poisonous substances that gain entry to the system. The manner in which the liver deals with such noxious substances as methylcholanthrene and phenobarbitol through enzyme neosyntheses has already been described (see p. 124), and it is known from studies with actinomycin D (Gelboin and Blackburn, 1963) and puromycin (Conney and Gilman, 1963) that these syntheses are gene-dependent. An obvious suggestion is that the steroid hormones are recognized by the liver not as hormones but as noxious substances and that they are dealt with accordingly.

If this is true then the disparity of action of the glucocorticoid hormones between the hepatic and extrahepatic tissues can be satisfactorily explained, and it follows that the main actions of these hormones, including the inhibition of protein synthesis, can be studied only in the extrahepatic tissues. The evidence from these tissues is still inadequate, but from studies of the way in which the glucocorticoid hormones cause involution of the thymus, Hechter and Halkerston (1965) have already concluded that cortisol may not act directly at the gene locus.

Finally, any consideration of the glucocorticoid hormones must take account of their role in support of chalone action (see p. 106).

CONCLUSIONS

It is evident that most if not all hormones are key factors in homeostatic mechanisms, which may operate at a variety of levels. Most commonly they act to modify certain metabolic processes in response to signals that originate externally to the animal, but they may also act to control animal populations by determining the timing and duration of breeding seasons and, if the ectohormones or pheromones are included, to control the structure of animal societies. Also, although the emphasis here is on animals, it is clear that similar mechanisms exist in plants, as when environmental changes lead to the production of a hormone which breaks the dormancy of buds (see Bonner, 1965).

HORMONES AND HOMEOSTATIC MECHANISMS

In an adult animal the impression is gained that each of the body tissues is set to operate around a particular functional norm, and that this is maintained by the basic homeostatic mechanism of which a chalone forms a part. In a target tissue this functional norm is low and the tissue only becomes fully functional when the basic mechanism is modified by a second homeostatic mechanism of which a hormone forms a part. Thus hormones operate by directing periodic variations from the norm.

Such variations are necessary to meet the sudden demands that are made on the target tissues as a result of changing circumstances, most of which originate outside the animal. Examples are numerous: in the mammals they include the way in which a sudden intake of sugar leads to the increased secretion of insulin, or a sudden stressful situation leads to the increased secretion of both adrenalin and a glucocorticoid hormone; in the insects they include the way in which, in the bug *Rhodnius*, a full meal results in the production of ecdysone which leads to moulting; and in the angiosperms they include the way in which changing day length results in the production of florigen which leads to flowering. These are phenomena ranging from the intermittent to the seasonal, of which the most dramatic are the hormonally controlled annual breeding cycles of the vertebrates (see Bullough, 1961). They are also all phenomena which depend ultimately on an external stimulus, but one exception to this usual rule is provided by the hormonally controlled mammalian oestrous cycles, the durations of which are determined by some internal control based in the genes (Grüneberg, 1952).

Thus the hormones form one link in a system of intercellular communication, and the problem they present ranges from their synthesis and release to the manner of response of the target tissues. The production and release of hormones may occur in response to chemical messages such as increased sugar concentration, to nervous messages as during stress, or even to the dictates of some internal gene-dominated clock as in oestrous cycles.

Two examples which illustrate all these activating mechanisms are provided by the insects (see Wigglesworth, 1965) and the mammals (see Bullough, 1961). In the insect *Rhodnius*, the ultimate cause of moulting is a full meal, which activates the abdominal stretch receptors so that a continuous series of nervous messages passes to the brain. Certain neurosecretory brain cells then produce a neurohormone which passes as a chemical message to the prothoracic glands, which respond by secreting ecdysone. This circulates throughout the body to initiate the chain of tissue reactions that constitute moulting.

In a similar way the initial stimulus to the sexual activation of a mammal may come from the sense organs. Thus the onset of the breeding season in a ferret is triggered by increasing day length, while in a sheep it is triggered by decreasing day length. These external changes are perceived by the eyes from which a series of nervous messages passes to the brain and so to the hypothalamus. However, the hypothalamus is also the seat of an internal physiological clock, which in blind animals or in artificial conditions of constant day length is capable of initiating an annual breeding season at approximately the right time. Normally the external and internal stimuli reinforce each other.

In the stimulated hypothalamus certain neurosecretory cells produce neurohumoral substances which pass through the portal vessels of the pituitary stalk into the anterior pituitary gland (see Fortier, 1963). This is a similar sequence of events to that in *Rhodnius*, and indeed Bern (1966) has emphasized that "the neurosecretory neurone is the *sine qua non* of neuroendocrine regulation, representing as it does 'the final common path' between the nervous system and the endocrine system". It is this neurosecretion that stimulates the production of the pituitary gonadotropic hormones, and these in their turn stimulate the gonads to secrete the oestrogens and androgens on which depend the growth and proper functioning of the accessory sexual structures and the secondary sexual characters. Part of this final stimulus is to the nervous system through which further messages then reach the brain to reinforce the activation of the hypothalamus. This is the most complex chain of hormonal communications known, involving as it does two nervous components, three separate hormone systems, and a reinforcing feedback mechanism.

Since so many hormonal homeostatic mechanisms depend for information on sense organs of one kind or another, it is not surprising that both in the insects and in the vertebrates the main endocrine control centre is attached to the brain.

RESPONSES IN TARGET TISSUES

All the various hormonal control mechanisms must have been established by evolutionary selection, and therefore they must be gene-dependent. However, it is clear that such evolution has more often depended on changes in the responding cells than on changes in the hormone molecules. In any major group, such as the vertebrates, the molecular structure of the hormones is remarkably uniform and "flexibility in adaptation has been secured by modification of the target tissues" (Barrington, 1964). This point was emphasized by Bullough (1946) in relation to the target tissues of both oestrogens and androgens. The androgens are known to induce the growth and specialization of the posterior region of the kidney of the stickleback, of the skin on the thumbs of the frog, and of the comb and wattles of the domestic fowl, and it is clear that these are all examples of evolutionary modifications in tissue responses. They do not depend on any modifications in the chemical structure of the hormone itself.

It seems probable that hormones may act on a variety of cell functions in their target tissues. Thus their action may sometimes be on cell membranes, as in the case of the anti-diuretic hormone of the neurohypophysis, or on enzyme systems in which the tertiary structure of one or more enzymes may be allosterically deformed. In the words of Monod *et al.* (1963): "The specificity of hormones; their capacity for simultaneously activating or inhibiting a variety of metabolic processes and of exerting different effects on different tissues; the surprisingly small number of reactions in which they have been proved to take part as reactants as opposed to the large number of enzymes upon which they have been found to act; the lack of chemical reactivity of certain hormones such as the steroids; all in fact of these physiologically essential and chemically bewildering properties could be accounted for by the assumption that hormones in general act as allosteric effectors, each of them able specifically to trigger allosteric transitions in a variety of different proteins". The main criticism of this attractive theory is that "it is able to explain nearly everything" (Karlson and Sekeris, 1966).

However, in the case of the only two hormones, ecdysone and oestradiol, that have been adequately studied the impression is gained that the action is mediated through the genes but only at second hand through some interference with an intermediate effector molecule. Here again the theory of allosteric deformation has been applied. It has been suggested that this effector molecule is a protein which in a target tissue is specifically deformed in the presence of the relevant hormone. This key protein could be the repressor postulated by Jacob and Monod (1963) or the tissue chalone described by Bullough and Laurence (1964) or some other constituent molecule of the chalone homeostatic mechanism.

In the case of the oestrogens and androgens it has been suggested that a primary action may be to interfere with the chalone mechanism to accelerate mitotic activity and so to force larger numbers of cells along the pathway to function and death. It has often been noted, especially in embryonic tissues, that a burst of mitosis commonly precedes tissue function. In adult mice Stockdale and Topper (1966) have shown that "mammary-gland epithelial cells must first divide in order to synthesize casein", while in both insects (Wigglesworth, 1965) and mites (Hughes, 1964) active epidermal mitosis and active cell loss accompany the production of a new cuticle at ecdysis. It appears that the cells of many target tissues do not progress far enough along the functional pathway for tissue function to be possible unless with an increased mitotic rate the cell flow to death is also increased.

ACTIONS OF PHEROMONES

As already mentioned, it is possible to extend this consideration of hormones to include another group of chemical messengers, the ecto-hormones or pheromones (see Karlson and Butenandt, 1959; Weaver, 1966). These also provide stimuli that originate externally since they are passed from individual to individual. In insects they include body secretions which are used for many purposes such as "protection, sexual excitation, sexual attraction, as stimuli to hormone production or possibly as hormone-like substances themselves" (Wigglesworth, 1965). Karlson and Butenandt (1959) have pointed out that they fall into two main groups, the telemones and the pheromones proper. The telemones induce a change in the behaviour of the recipient, as when a female moth secretes an odoriferous substance which attracts the male (see Schneider, 1966). The pheromones proper induce a major change of function in the recipient, as when a queen bee exudes the "queen bee substance" (9-oxodecenoic acid) which suppresses ovarian growth in the worker females. There is probably no clear distinction between telemones and pheromones, and indeed the "queen bee substance" also acts as a telemone to attract the drone during the mating flight.

Other examples of pheromone action are provided by the termites. The reproductive castes are believed to give off substances to prevent the development of any other reproductive individuals, while the presence of sufficient numbers of soldiers tends to prevent the production of any more. None of these reactions has been analysed at the cellular level but it seems that, either directly or more probably indirectly, at least some of the pheromones must influence gene activity.

THE HIERARCHY OF GENE CONTROL

The general impression emerging at this point is of an originally totipotent genome being acted on and controlled throughout the whole life of an individual multicellular animal by a sequence, or hierarchy, of chemical messengers. First, there are the repressor molecules produced by regulator genes. Second, there are the inducing agents which operate sequentially in an appropriate pattern to limit the genetic potentialities of the embryonic cells and thus to create the tissues and the organs and the general morphology of the adult. Third, there are the chalones which operate in the adult in those tissues that retain the alternative genetic potentialities of mitosis and of ageing with tissue function. The main task of these various chemical signals, which together may make up the whole language of gene control, is to create a mature organism within which, in all the constituent tissues, cell replacement and function and death are appropriately balanced.

However, the ability of this mature organism to adapt itself appropriately to the challenge of environmental change depends to a large extent on yet another system of chemical messengers, the hormones. These may interfere in cell metabolism at a variety of levels, but increasingly it is being realized that many of them influence gene action. Present evidence suggests that this influence is not exerted directly on the genes: with adrenalin and hydrocortisone the action may be through the intermediary of the chalone mechanism, and with both ecdysone and oestradiol "an intermediate control system" is needed to link their actions to the genetic loci (Hechter and Halkerston, 1965). The hormonal homeostatic mechanisms may thus be set above the tissue homeostatic mechanisms, which are themselves responsible for gene control. Many of the pheromone systems may be similar.

Thus the hierarchy of gene control in a metazoan includes the repressor molecules, the inducing agents, the chalones, the hormones and the pheromones, and of these the hormonal and pheromonal mechanisms are obviously the most recently evolved. It is through them that an animal is able to respond adequately to a range of situations which in the evolutionary history of the species have proved to be of particular significance.

The conclusion is that throughout the whole of life the control of differentiation and thus of cell function is maintained by means of a constant stream of instructions, which altogether constitute the language to which the genes respond. All these instructions are in the form of changing concentrations of chemical messengers, some of which must act directly at the gene level while the others act only indirectly. However, although all tissues are formed by and react to instructions of these kinds, certain adult tissues have lost their capacity to respond. Such tissues, in which alternative genetic syntheses are no longer possible, may be common in short-lived adult insects, while in adult mammals they include the two major systems, the nerves and the striped and cardiac muscles.

CHAPTER 7

Carcinogenesis

It now appears probable that all metazoan tissues that retain the capacity for mitosis are dominated by tissue-specific homeostatic mechanisms within which tissue-specific chalones evidently act as chemical messengers. It is also probable that the various homeostatic mechanisms are all built on exactly the same plan: all of them act to maintain a proper balance between mitotic activity on the one hand and ageing and tissue function on the other. These mechanisms must obviously be liable to damage, both inherited and acquired, and if as a result there is a significant alteration in the balance between cell production and cell loss, then a local overgrowth, or cancer, could be one of the possible consequences. It is therefore obvious that an understanding of normal tissue homeostasis could provide an invaluable foundation on which to build a better understanding of the cancer problem.

In recent years two important points of agreement have begun to emerge, namely that carcinogenesis is the outcome of a local breakdown in tissue homeostasis, and that this breakdown involves changes in gene activity. The problem is, in fact, one "of the impact, upon the genomal integrity, of any of a host of reagents" (Haddow, 1964). The first of these reagents to be studied intensively were the chemical carcinogens, the next were certain physical factors, especially ionizing radiations, while most recently attention has been turned to the carcinogenic viruses (see Hieger, 1961; Ambrose and Roe, 1966). It is probable that the primary intracellular actions of these agents are diverse, but it is clear that the secondary actions must all converge to produce a similar kind of damage within the affected cells. Thus it is not the agents themselves but the damage that they ultimately cause that is of critical importance.

There are at the moment two main theories: first, that the essential damage is extrachromosomal, and second, that it is genetic. Cancer is in fact such a diverse phenomenon that there is no reason why both theories should not be correct. The important point may be that in both cases gene activity is profoundly and permanently modified.

The Process of Carcinogenesis

One difficulty in approaching the problem of cancer lies in the wide range of pathological conditions that are included under this name. It has been said that the word cancer has as wide a content of meaning as has the word infection. However, just as there are common features in all processes of infection, so also there must be common features in all forms of carcinogenesis. Some of the more basic of these were recognized by Berenblum (1941) and Berenblum and Shubik (1947) in mice, and by Rous and Kidd (1941) in rabbits. More recently they have been analysed in detail by Foulds (1963, 1964, 1965), and it is now possible to distinguish clearly between the three separate processes of initiation, promotion, and progression. It is during initiation that the cancer cells are formed, whether "spontaneously" or by some kind of carcinogenic agent; it is by promotion that these cancer cells, which may have lain dormant for weeks or months or years, are stimulated to multiply to form a visible tumour; and it is by progression that the tumour cells tend to become ever more malignant.

INITIATION

As already mentioned, it is now widely believed that carcinogenesis depends essentially on the modification of gene action either as a result of extrachromosomal damage or of some somatic gene mutation. Since in the following argument it matters little which of these is correct, the theory of gene mutation is accepted for convenience as a working hypothesis. Certainly this theory agrees with the facts that "initiating action is brought about very quickly, perhaps even instantaneously; it is irreversible; and it appears to be highly specific" (Berenblum, 1954).

It must, however, be mentioned that strong objections have been raised to this theory of somatic mutation. Rous (1959) has emphasized how little evidence actually exists, and Hieger (1959, 1961) has objected that gene damage must be a comparatively rare event and that if, for cancer to develop, several distinct mutations are needed, as indeed seems to be the case, then the chances of initiation decrease "to practically zero". Burnet (1957, 1962) has calculated that "at any given genetic locus an error in replication occurs with a frequency in the range $10^{-15} - 10^{-7}$ per replication", but he has also pointed out that in an animal as large as man some 10^{14} cells may be constantly reproducing themselves.

However, all such arguments omit a number of important considerations. First, they do not take into account the effects of carcinogenic agents in increasing the mutation rate (see Lamerton, 1964), and it must be emphasized that "spontaneous" tumours may in fact be induced by as yet undetected carcinogens. It is, incidentally, not essential that carcinogens should themselves be mutagens, since the gene damage could occur as the end-effect of a series of repercussions within the affected cells.

Second, such arguments do not allow for the fact that gene damage may occur in non-dividing cells, although Curtis (1963) has presented evidence that certain non-dividing mammalian cells accumulate deleterious mutations at such a rate that by late middle age "virtually all cells carry many gene mutations". This damage is visible, like that which is known to be typical of chronic myeloid leukaemia (Brown and Tough, 1963), but much critically important gene damage may be invisible. Some could even occur "spontaneously" as the result of the decay of C^{14} within the DNA molecules (Rytömaa, 1967).

Third, the arguments against the mutation theory of cancer do not take account of the possibility that part of the necessary gene damage may be inherited, although in man Burch (1962, 1963) has suggested that the majority of malignancies may be at least partly dependent on this. It is this that evidently leads to the existence of cancer-prone families (see Lynch et al., 1966), within which, however, carcinogenesis is not confined to a single organ or tissue. This agrees with the conclusion (see p. 118) that much of the tissue homeostatic mechanism, and especially that which controls the balance between cell production and cell loss, must be common to all tissues.

Initiation, then, may depend on a specific pattern of damage to those genes that are involved in cellular homeostasis, and this damage may be caused directly or indirectly by chemical or physical factors, or by viruses. With the viruses there is also the possibility that carcinogenesis occurs not because of gene mutation but because the instructions issued by the viral DNA or RNA conflict with those of the host cell in such a way as to disrupt the homeostatic balance (Kaplan, 1964; Lwoff, 1966).

However, "initiation evokes no distinctive change . . . and, in particular, it incites no proliferation of cells" (Foulds, 1963). The impression is that the damaged cells remain under the control of the tissue homeostatic mechanism, perhaps because of the influence exerted by the surrounding normal cells. Some further event is clearly needed to release them from this control. However, it is by no means certain that their release will ever occur, and it is probable that any older mammal, dying for other reasons, contains numbers of dormant cancer cells.

Theoretically, too, a number of other fates may overtake these cells. First, if they form part of an epithelium, such as epidermis or duodenal lining, they may be carried with the flow of cells to the surface and there shed. Second, it now appears that many, if not all, cells contain some mechanism whereby damage to the DNA can be repaired if it is not too severe (Dean *et al.*, 1966). Third, if the damaged genes begin to produce new, or "foreign", types of protein these may act as antigens to provoke an immune reaction, which may destroy the affected cells. It follows that any tumours that do appear are unlikely to exert any strong antigenic activity, and this seems to be true (see Alexander, 1966).

PROMOTION

If a dormant tumour cell avoids all these fates, then at some time it may receive the stimulus to multiply and thus to produce a visible tumour which then grows progressively. This is promotion and the agent providing the stimulus is known as a promoting agent. It must be stressed that typical promoting agents, of which croton oil has been most widely used experimentally, are not ordinarily carcinogens. Consequently initiation must precede promotion, and if experimentally promotion precedes initiation then tumours may not be produced. Evidently the process of promotion is quite different from that of initiation.

The nature of promotion has been closely examined by Berenblum (1954), who points out that, unlike initiation, it "is slowly acting; it is, in some degree, reversible; and it is less specific". He discusses the possibility that the change from dormancy to progressive growth may depend merely on a simple non-specific stimulus to cell division such as is seen in wounds. There is an old belief that, given sufficient time, irritation or ulceration will automatically lead to hyperplasia and cancer, and Berenblum (1941) in his original and now classical experiments used the irritating substance croton oil which causes a pronounced epidermal hyperplasia. Wounding can also act as a promoting agent in carcinogen-treated skin, while in the liver the regeneration that follows partial hepatectomy or accompanies cirrhosis, promotes the earlier appearance of liver tumours (Glinos et al., 1951; Laws, 1959, 1960). However, both wound healing and regeneration tend to be short-lived episodes, and a stronger promoting action is obtained by repeated tissue damage (Pullinger, 1945).

Tumours may also form in embryos and young animals, and in this case promotion and subsequent growth are particularly rapid "because the normal expansive growth, characteristic of that stage of development, itself serves as the promoting stimulus" (Berenblum, 1954).

In spite of this and much similar evidence, Berenblum (1954) has concluded that a non-specific stimulus of cell division is inadequate to explain promotion. Although obviously the dormant cancer cells must undergo repeated divisions during promotion, he insists "that promoting action is essentially a process of delayed maturation". In other words a promoting agent dislocates the normal balance that exists between cell gain on the one hand and cell maturation and death on the other. Certainly this must be an essential feature of any tumour, but since, from what has been said earlier, such dislocation is probably genetic in nature, it is perhaps more likely to result from the damage sustained during initiation. The act of promotion may merely serve to demonstrate its existence. Thus it still remains difficult to determine the relative importance in promotion of the increased mitotic rate and of the delayed maturation and death of the cells.

Yet another type of promotion is found in hormone-dependent tissues. Here the stimulus to growth is provided by the hormone and such tumours continue to grow so long as the hormone is present, regress when it is withdrawn, and recur at the same place and with the same characters when it is restored. This type of reaction has been well analysed by Foulds (1963) in relation to the mammary glands of mice, in which the earliest stages of visible cancer are recognizable as "nodules" or "plaques", which grow during the second half of pregnancy but regress between pregnancies. When they are regressed "fibrotic remnants containing atrophic tubules lined by ragged epithelium are sometimes present but often only slight traces or none are discoverable. Nevertheless in mice allowed to live, the plaques grow again, at the same place as before, during the second half of the next pregnancy, however long it is delayed". These cycles of growth and regression may continue for some time without further event but ultimately one of the plaques may change its behaviour and from "a pregnancy-

F

dependent plaque, a pregnancy-independent tumour is established".

Emphasis has often been placed on the distinction between the irreversible type of promotion, caused for instance by croton oil, and the reversible type of promotion which is conditional on the presence of a particular hormone. However, there is probably no valid distinction. Promotion is evidently what happens when a group of cells, damaged during initiation, escapes from the growth-restraining influences that have previously kept it quiescent, and the degree of freedom achieved varies in different cases. In normal tissues it is commonly complete or it would not show at all; in hormone-dependent tissues, while it remains marginal, growth can only occur in the presence of the hormone.

In considering tissue homeostasis it was emphasized that normal tissue cells continue to undergo mitosis not because of any stimulus to do so but in the absence of a signal to stop. Similarly, regarding cancer cells, Foulds (1963) has concluded that "neoplastic growth is attributable more probably to the failure of a normal repression than to a positive stimulation of cell division". As already shown, the only natural antimitotic signal known is that carried by the tissue chalone. It may therefore be suggested that one characteristic of the dormant tumour cells is a failure to synthesize, or a failure to retain within themselves, sufficient chalone for adequate self-control. The creation of this situation may be an integral part of initiation.

However, so long as such cells exist singly or form only a very small cluster, they may be expected to remain dormant because they may receive adequate amounts of chalone from the surrounding cells. Then, as Kaplan (1964) has said, "as the tumor grows in size . . . unresponsiveness to local growth-restraining influences will tend to increase". Thus the creation of a favourable environment for cell multiplication within the invisible and dormant micro-tumours may be an integral feature of promotion. Promotion is complete when the cell mass has acquired a radius greater than the effective length of the chalone diffusion path. From a study of chalone lack in the edges of wounds Bullough (1965) has suggested that in epidermis "this critical cell mass might have a diameter of a little less than 1 mm".

In a hormone-dependent tissue the role of the hormone as a promoting agent is only slightly different. Evidently the normal condition of a hormone-dependent tissue is one of almost complete mitotic inertia due to the high chalone content of the cells, and in such a tissue the chances of promotion of any dormant tumour cells are negligible. In the case of the mammary glands it has been suggested, following the evidence of Prop (1965; see p. 139), that the mitogenic hormones may act by neutralizing the tissue chalone. In this situation any dormant tumour cells that synthesize or retain slightly less than the normal amounts of chalone may still possess enough to prevent growth when the hormones are not present, but may be expected to show a higher than normal growth response during pregnancy and lactation. When the hormones are withdrawn the chalone content will again be adequate to suppress growth.

PROGRESSION

The final step, or more accurately the final series of steps, in carcinogenesis involves a further gradual breakdown in the mitotic and functional homeostatic mechanisms, which usually leads to faster and more independent growth. When the tumour cells emerge from dormancy it is probable that the condition of their homeostatic mechanisms is not significantly altered. It is merely the size of the group that allows the cells to break free from tissue control and to multiply autonomously. However, from this time onwards their characteristics, whatever they may be, tend to change progressively. The characteristics of such progression are: that the changes are haphazard both in their timing and in their character; that they do not occur in all the existing tumour cells but only in a small fraction, and sometimes perhaps only in one cell; that they occur independently in different tumours in the same animal or in different areas of the same tumour; that they are independent of the size or age of the tumour; and that they occur in abrupt steps and are irreversible (Foulds, 1964).

As Rous and Kidd (1941) have emphasized, what happens at each step is no mere exaggeration of the previous state "but a wholly new event", the genesis of a new type of cancer cell that is quite distinct from the one from which it arose. This is the type of change that would be expected if the underlying cause was progressive mutation (Klein and Klein, 1957), again with the proviso "that the term mutation is used in its broadest sense, simply meaning stable irreversible hereditable changes without regard to their intra- or extrachromosomal nature". (Foulds, 1963).

Despite its extreme complexity, progression typically involves only three changes of major importance. The first, which is the least important, is the progressive loss of ability to complete the normal tissue syntheses. The other two, which are critically important, lead to a progressive increase in the mitotic rate and to a progressive change in the state of the cell surface. Any changes that occur are of significance only if they enable the altered cells to prosper in competition with the other tumour cells. By a process of cellular "selection of the fittest", those cells survive and produce more descendants which have the highest mitotic rate, or which, because of the abnormality of their cell wall, are able to detach themselves from their tissue of origin and migrate to colonize other parts of the body.

It can also be suggested that the increasing mitotic rate during progression is all that is required to depress normal tissue syntheses, since cells that are committed to mitosis have no opportunity for any other type of synthesis (Kaplan, 1964). It would be interesting and important to discover whether such cells could ever revert to normal tissue syntheses if they received the correct signal to do so.

It is important to note that these changes in tissue function, mitotic rate, and cell surface do not necessarily progress with equal speed. Thus a tumour with a high mitotic rate may have cells with such normal cell walls that no migration, or metastasis, occurs, while a tumour which is hardly growing at all may have cells with such abnormal cell walls that metastases are widespread.

Theoretically the ultimate step in progression would produce cells which showed no trace of the characteristics of their tissue of origin and which, if the blood supply remained adequate, were all involved in rapid mitotic cycles. Such a state is not usually seen while the tumour is in the animal in which it originated; usually the animal is killed by the tumour before this degree of progression is attained. However, tumour cells which have reached or are close to this ultimate stage can be obtained by transplanting the tumour from host to host. Many ascites tumours are typical examples and in them the mitotic cycle may be completed in some 11·5 hr (Defendi and Manson, 1963), which is close to the theoretical minimum.

One important rider must be added to this account of progression. Rarely, instead of the tumour cells changing from benign to malignant behaviour, the reverse occurs. As Smithers (1967) has said, "tumour behaviour in man is certainly not an invariable, one-directional, fixed performance . . . conforming to a pre-determined pattern and beyond all control". Although there is indeed "a general tendency towards progressive escape from control", the occasional tumour changes its behaviour pattern "so drastically as to disappear altogether".

CHALONES AND CARCINOGENESIS

From this brief review it is evident that the cellular breakdown which leads to cancer is essentially genetic. With increasing age gene damage may tend to accumulate in most if not all tissues, and if in any cell the pattern of gene damage is appropriate then carcinogenesis may be initiated. The development of this appropriate pattern may be hastened by carcinogens and augmented by inherited gene weaknesses. One essential feature is that this damage must lead to a disruption of the normal balance between cell production and cell loss.

Attention may now be turned to the possible ways in which gene damage may affect tissue homeostasis (see Bullough, 1964; Kaplan, 1964), and it may be confidently expected that one common feature in carcinogenesis will prove to be a failure in the system of intra- and intercellular chemical messengers on which homeostasis depends. In particular the tissue chalone which controls the normal mitotic rate may be involved.

THE BREAKDOWN OF TISSUE HOMEOSTASIS

In adult tissue cells most of the genes are blocked, apparently permanently, and on the assumption that gene damage is random, much of it must be confined to these genes. Such damage should have no visible effect unless it leads to the reactivation of the blocked genes, which in fact never seems to happen. Any abnormal reopening of genes closed during embryology would be expected, at least sometimes, to result in teratomata, but such growths seem to originate only from germ cells in which, of course, the entire genome remains potentially functional.

Damage might also occur in those genes which specify essential metabolic enzymes, but this would merely lead to cell weakness or death; or in those genes which direct tissue function, but this would only marginally reduce the efficiency of the tissue; or in those genes which direct the process of mitosis, but this might merely lead to the death of the cell when it next attempted to divide.

To be of importance in carcinogenesis the damage must involve those genes which, by initiating mitosis and controlling cellular ageing, determine the numbers of cells in, and therefore the mass of, the tissue or organ. In any tumour it is primarily cellular homeostasis which has been disrupted; if the tissue genes are also affected this may determine the appearance of the tumour, and so enable it to be named and classified, but modified tissue function does not itself contribute to the actual formation of the tumour.

In other words the only significant damage in carcinogenesis is that which confers a proliferative advantage on the cells, and this can only come from an increased mitotic rate which is not offset by an increased rate of ageing, or from a decreased rate of ageing which is not offset by a decreased mitotic rate, or from both. One critical characteristic of a carcinogen may be that it tends to inflict precisely this type of mutational damage (Trainin *et al.*, 1964).

The degree of proliferative advantage gained by the damaged cells must obviously vary according to the numbers and the localities of the mutations that have occurred and also according to whether these mutations are partial or complete; the consequences to cell morphology must depend on the degree to which the tissue genes are also involved. Partial gene failure could result in an inadequate supply of some particular molecule, either because its rate of synthesis is reduced or because its structure is imperfect; total failure may result either in no synthesis or in the synthesis of an entirely abnormal molecule.

The possible outline of the mechanism of tissue homeostasis is discussed on p. 121 and summarized in Fig. 25. Within this scheme there are several obvious locations at which genetic damage could have serious consequences in upsetting the normal balance between cell gain and cell loss.

Theoretically a tumour should arise instantaneously and grow uncontrollably if the genes of the "mitosis operon" fail to respond to the antimitotic message of the tissue chalone. Such immediate tumours, growing without the intervention of a dormant period, do sometimes occur but they are relatively rare.

A second possibility is a failure in one or more of the genes that are involved in the chain reaction of ageing. This should lead to the "delayed maturation" that Berenblum (1954) has stressed must be basic to all forms of cancer, but in a tissue with a relatively low mitotic rate there may be no immediate effects. The cells, with their mitoses controlled either by endogenous or exogenous chalone, should enter a dormant period.

After the dormant period the rate of growth will depend on whether other types of genetic damage have also been sustained. If, for instance, the cells synthesize inadequate amounts of chalone then the mitotic rate should rise in proportion, and the same result should also be obtained if the cells fail to maintain a cell wall which is adequate to prevent undue chalone loss. Changes in the structure of the cell wall could also result in the failure of the cell to maintain its position in the tissue, so that metastasis occurs.

If the damage to the genes is only partial then theoretically the growth period that follows promotion should be only temporary. The ultimate mass of the tumour, like the ultimate mass of a normal tissue, must depend on the balance between cell production and cell loss, and during its growth period a tumour may merely be moving from one point of balance to another. However, owing to progression, any point of balance that is reached may be only temporary, and when, at the theoretical limit, the damage to the genes is total and all cells remain in the mitotic cycle, the point of balance recedes to infinity.

CHALONES AND DORMANCY

It appears that during dormancy a tumour cell remains submissive to the tissue homeostatic mechanism, and since the majority of tumours are believed to pass through this phase it is particularly important to understand the nature of the controlling influence, which is probably a chemical messenger. One obvious suggestion is that a dormant tumour cell continues to respond to the antimitotic message of the tissue chalone coming from the surrounding normal cells.

Theoretically dormancy may be broken if the cell suffers a further mutation which impairs its ability to respond to the inhibiting influence of the surrounding cells. Alternatively dormancy may be broken if a local decrease in the chalone concentration leads to a raised mitotic rate and so to the creation of a tumour of such a size (about 1 mm diameter; Bullough, 1965) that its central cells are beyond the inhibiting influence of the surrounding cells. Since it is generally agreed that promotion does not usually involve mutation, the second alternative is the more likely.

Thus in any tissue the chance that dormancy will be broken may seem to be in direct proportion to the mitotic rate, which is itself inversely related to the strength of the chalone mechanism. However, account must also be taken of the rate of cell loss. Any tissue, such as duodenal mucosa, with a naturally high mitotic rate also shows a naturally high rate of cell loss. Curtis (1963) has stressed that such tissues show the least signs of chromosome damage, and with damaged cells most likely to be thrown off, the chances for carcinogenesis are low.

The actual chances of promotion may therefore depend on the rate of mitosis *minus* the rate of cell loss. From this it seems that the chances should be equally poor in tissues with either a low or a high mitotic rate, and that it is those tissues with an intermediate mitotic rate, or with an oscillating mitotic rate, that should be most susceptible. In such tissues, as Smithers (1967) has said, "neoplastic disease occurs most frequently at sites where there is the greatest demand for repair", and it is "recurrent rather than continuous . . . growth (that) is by far the commonest precursor of a disorganized growth pattern in man". It is exactly this type of intermittent growth, whether induced by wounding or by mitogenic hormones, that is believed to depend on a weakened chalone mechanism.

STRESS HORMONES AND DORMANCY

If the chance that the period of dormancy will be short depends on a high rate of cell production which is not offset by a high rate of cell death, then the converse should also hold good. Anything that strengthens the chalone mechanism and so depresses the mitotic rate of the dormant cells should tend to lengthen the period of dormancy. At the limit, the average duration of dormancy might be prolonged beyond the normal length of life of the animal and then cancer, though present, should never appear.

It has been shown (p. 105), at least in some tissues, that the chalone operates more efficiently in the presence of adrenalin and that this joint action is still further strengthened in the presence of a glucocorticoid hormone. Thus it may be expected that the two stress hormones will act as retarding agents to prolong dormancy and so to inhibit carcinogenesis. Unfortunately the literature is confused by the many studies on stress in which no proper distinction has been made between initiation, promotion, and subsequent tumour growth. However, it has been clearly shown that dormancy is prolonged in mice that are stressed by high temperatures (Fuller et al., 1941; Wallace et al., 1944, 1945) or by excessive noise (Molomut et al., 1963), and Anderson (1964) has drawn the general conclusion that "apprehension (gives) some protection against carcinogenesis".

It is difficult to devise stressful situations in which the degree of stress inflicted can be accurately varied and in which the animals themselves are not damaged. However, in the wild the commonest and most natural form of stress is probably hunger, and the experimental difficulty of stressing animals has been most successfully circumvented by the use of restricted diets. The stress of hunger is known to lead to enlarged adrenal glands (Boutwell et al., 1948; Sayers, 1950), to the increased secretion of both stress hormones (Galicich et al., 1963; Godefroy, 1964), and to a reduced mitotic rate (Bullough and Eisa, 1950). The most extensive and important studies on the effect of hunger on carcinogenesis are those of Tannenbaum (1940, 1942, 1947) and of Rusch et al. (1945), although none of this work was originally considered in terms of stress. It was Tannenbaum who first showed that in a wide variety of dormant tumour cells, both spontaneous and induced, promotion can be delayed and even prevented in mice maintained on a diet of two-thirds of what they would naturally eat. He also showed that such animals remained in excellent health and lived longer than did the well fed controls.

White (1961) has surveyed the now extensive literature on restricted diets and carcinogenesis as well as the many unsuccessful attempts that have been made to explain the delayed appearance of the tumours in terms of the diets themselves. It now seems certain that the effect is in fact due to stress (Bullough, 1965).

There is also a large literature on the effects of the glucocorticoid stress hormone on carcinogenesis, although the effects of adrenalin have not yet been studied. Again some confusion has been caused by the many failures to distinguish between initiation, promotion, and subsequent tumour growth, but a particularly well planned series of experiments has been described by Trainin (1963). By studying carcinogenesis in mouse skin he showed that after hydrocortisone treatment or after adrenalectomy the rate of initiation was unchanged but the rate of promotion was strikingly altered. Against a control tumour yield of 77%, the yield after hydrocortisone treatment was reduced to 11%, while after adrenalectomy it was increased to 100%. Similar results have been obtained with spontaneous leukaemias (Upton and Furth, 1953; Wooley and Peters, 1953; Jaffe *et al.*, 1963) and with induced lung tumours (Gillman *et al.*, 1956). Law (1947) has also shown that adrenalectomy increases the incidence of spontaneous lymphoid leukaemia in mice, while Metcalf (1960) has found that AKR mice are naturally characterized by a low glucocorticoid hormone production and a high rate of spontaneous leukaemia. The fact that adrenalectomy increases the mitotic rate and eliminates the diurnal mitotic cycle was shown by Bullough and Laurence (1961).

The conclusion is clear that the adrenal glands, through the actions of the stress hormones, act to retard promotion, and the suggestion is that they achieve this by strengthening the various tissue chalone mechanisms and so reducing the mitotic rate. In view of the obvious practical importance of retarding agents it is curious how little attention has been paid to them, and this is especially so in view of the considerable attention that has been given to promoting agents.

CHALONES AND TUMOUR GROWTH

It thus appears that promotion usually depends on changes in the environment of the dormant tumour cells and not on changes in the cells themselves. If this is true then, at least in the early stages of tumour growth, the cells should still be capable of responding to changes in the tissue homeostatic mechanism, and especially in the chalone concentration. The available evidence suggests that this is so. The effects of chalone lack are demonstrated when small adenomatous hepatic nodules, implanted subcutaneously, are able to grow more rapidly following partial hepatectomy, although with increasing malignancy this response is no longer seen (Trotter, 1961).

Conversely, tumour growth may be delayed when the chalone mechanism is strengthened by the presence of the stress hormones. This leads to the persistence of a diurnal mitotic rhythm in some tumours, and here again as malignancy increases the rhythm is lost (see Bullough, 1965). Stressful situations of various kinds have also been described as delaying tumour growth (Rusch and Kline, 1944; Rashkis, 1952; Marsh et al., 1959), and Pearson (1959) has indicated that when tumour growth is inhibited by starvation stress the cells may reacquire some degree of functional activity. This is what might be expected of cells that were forced from the mitotic cycle into the ageing pathway.

The glucocorticoid hormones have also been found to reduce the rate of growth in a lymphosarcoma (Heilman and Kendall, 1944; Stoerk, 1950), in lymphoid leukaemia (Higgins and Woods, 1950; Gabourel and Aronov, 1962), and in a variety of other tumours (Sugiura et al., 1950). However, with progression glucocorticoid sensitivity gives place to glucocorticoid independence, which again suggests that with increasing malignancy the effectiveness of the chalone mechanism is reduced. It is obviously important to discover whether this reduced effectiveness may be related to a reduced chalone concentration or to the reduced responsiveness of the cells to whatever chalone is present.

The evidence is unfortunately still limited, but apart from a suggestion that ascites hepatoma cells may lack liver chalone but still be capable of responding to it (Otsuka and Terayama, 1966), there have been two direct attempts to discover whether malignant tumours obtained by years of transplantation from animal to animal are still capable of responding to their respective tissue chalones.

In the first, Bullough and Laurence (unpublished) have studied the V × 2 epidermal tumour, which grows and metastasizes so rapidly that one small inoculation can kill a full grown rabbit within a month. It was found that the cells, which retain no visible trace of their epidermal origin, still synthesize the epidermal chalone, although possibly not in normal quantities. In a rabbit with a large tumour so much chalone is passed into the blood that the normal epidermal mitotic activity is almost completely inhibited. Even more important it was found, both in vivo and in vitro, that the tumour cells are capable of responding by mitotic inhibition to the presence of the epidermal chalone, but only if adequate amounts of the two stress hormones are also present.

The only obvious explanation of this situation is that the cells, besides losing their ability to age normally, do not contain an adequate concentration of the epidermal chalone to prevent mitosis. This could be due to a reduced ability for chalone synthesis or to a reduced ability to retain whatever chalone is synthesized. Certainly the abnormality of the cell wall is indicated by the high rate of metastasis.

The other study was made by Rytömaa and Kiviniemi (1967, and personal communication) using a transplanted granulocytic chloro-leukaemia of rats. Normal mature granulocytes synthesize a granulo-cytic chalone which passes in the blood to suppress mitotic activity in the granulocytic precursor cells in the bone marrow. By means of an *in vitro* technique it was found, first, that chloroleukaemic cells possess about $\frac{1}{40}$ of the normal granulocyte chalone content; second, that in chloroleukaemia the serum contains more chalone than does normal serum; and third, that the chloroleukaemic cells are capable of responding by mitotic depression to an adequate chalone concen-tration.

As with the V × 2 tumour, the implications are that the leukaemic cells continue to synthesize their chalone, although perhaps not in normal amounts; that this chalone escapes at an abnormal rate through the cell wall to produce the abnormally high serum levels; and that to suppress mitosis *in vivo* it is necessary to raise the chalone content of the whole body in order also to raise it in the leaking cells. With neither type of tumour has the critical experiment been performed to discover whether, with mitosis inhibited, the cells are able to enter the ageing pathway and perhaps even to activate the tissue genes.

Although the similarity in the reactions of these two different tumours is striking, the evidence is obviously inadequate to permit any wide generalization. However, it is now clear that even cells with a high degree of malignancy obtained by repeated transplantation may retain at least some ability to synthesize their tissue chalone and at least some ability to respond to it.

CONCLUSIONS

To the pathologist the clinical picture of cancer may be complex in the extreme, but this complexity evidently stems from the widely varying degrees of involvement of the tissue genes which do not seem to be of primary importance in carcinogenesis. Clearly the fundamental damage that leads to tumour formation is to the mechanism of cellular homeo-stasis, and this is a mechanism which is evidently common to all mitotic tissues. The evidence suggests that carcinogenesis depends on a specific pattern of mutations in the genes which control mitotic activity and cellular ageing. Some of these mutations may be inherited while others are acquired during life. With the acquisition of the last mutation the pattern is complete and initiation occurs. This genetic damage must lead to a dislocation of the normal balance between the rate of cell production and the rate of cell loss; it may also commonly lead to a reduced rate of chalone synthesis or an increased rate of chalone loss.

The important point is that the pattern of mutation required to initiate carcinogeneses may be the same in all tissues.

The initiated cells then usually lie dormant under the control of their normal neighbours, and it seems that the tissue chalone is partly or solely responsible for exercising this control. Promotion occurs when for one reason or another the dormant cells break free from this control. Carcinogenesis therefore also depends on the existence of an appropriate cellular environment. With the passage of time progression then commonly leads to increasing malignancy, but in the only two malignant tumours so far studied the cells are still capable of manufacturing and responding to the tissue chalone.

One common misconception is that tumour cells are dedifferentiated and even that they are reverting towards the embryonic condition. It has been stressed that a typical mitotic tissue is so far differentiated that only two alternative states of differentiation remain possible, the one resulting in mitosis and the other in ageing and tissue function. In many tumours both these possibilities still remain, although a new point of balance is established between them. In a malignant tumour all trace of tissue function may have been lost and all the cells may be involved in rapid mitosis, but even on the assumption that the loss of tissue function is permanent, this merely means that the cells are limited to differentiation for mitosis. In a non-mitotic tissue the cells are similarly confined to only one state of differentiation, in this case for ageing and tissue function. Obviously in neither of these cases has any previously closed part of the genome been reopened, as would be the most characteristic feature of dedifferentiation towards the embryonic condition.

BALANCED CELL POPULATIONS

The mass of any tissue or organ is determined by the relation between the rate of cell production and the rate of cell loss, and if either or both of these rates change then a new point of balance will be established which will be represented by a new cell mass. This is, in fact, what must happen in carcinogenesis when the mitotic rate rises, or the ageing rate falls, or both.

The most impressive feature of a tumour is its growth. However, in most tumours, growth may be characteristic only of that period when the cell mass is expanding towards a new point of balance. On a different level, it may resemble the growth of a regenerating liver fragment as it too expands towards the normal liver mass.

In many tumours the new point of balance is quickly reached and a small, "static, encapsulated, harmless swelling" is the result (Smithers, 1967); in others the dislocation may be so great that the new point of balance may not be reached before the tumour becomes so large that the animal is killed; in yet others the new point of balance may be a receding goal as the cellular homeostatic mechanism is further dislocated during progression.

Smithers (1967) in particular has emphasized that tumour growth has a common tendency to slow down as the tumour mass increases, and that this cannot be entirely due to a failure of the blood supply. Goodman (1957) has also emphasized the growth deterrent effect of increased tumour mass by showing that the growth rate of a mammary tumour may decelerate in the presence of a second tumour and accelerate after the removal of that tumour, while Schatten (1958) has found that the removal of a primary tumour may cause the more rapid growth of the metastases.

These results support the suggestion that tumour growth is not unlimited, and that in fact it continues only until cell production is once more balanced against cell loss. They also show that tumour mass is not determined solely by factors within the tumour cells. As with the mass of any normal organ, the ultimate mass of a tumour is partly a function of the total mass of the body within which it lies. When one tumour is removed another tumour at a distance may sense its loss in exactly the same way as when one kidney is removed and the other kidney senses its loss. In both cases the implication is that the removal of a mass of chalone-producing cells results in a fall in the chalone concentration in the whole body and therefore also in the surviving fraction of the tumour or organ. This means that any tumour which ceases to grow must still be producing the tissue chalone and must still be responding to it, even though the whole chalone mechanism must be operating on a lower than normal level of efficiency.

THE IMMUNE REACTION

The fact that gene damage increases with age has already been stressed (see p. 124). Although the types of somatic mutation that accumulate in this way are of differing degrees of severity and significance, clearly their steady accumulation must be most undesirable in any animal that is genetically programmed for a long life. Particularly dangerous must be the type of genetic damage which, given the appropriate cellular environment, leads to carcinogenesis.

One obvious feature of any cell that has suffered a somatic mutation is its uniqueness within the body, and this may express itself by the production of antigens which in turn may lead to the induction of antibodies. This has led to the argument that the primary function of the immune reaction may be the protection of the body against the dangers of somatic mutation (see Thomas, 1959; Prehn, 1960; Burnet, 1964). According to this view the wandering cells of the lymphocytic system may have been elaborated for the primary purpose of conducting a constant cell surveillance. They are cells that are specialized to seek out and destroy the cells of any tissue which, as a result of mutation, have produced cell surface components of a nature or of a pattern which is unnatural to that individual animal.

However, such a system, although invaluable, is obviously not perfect. It can only operate effectively if the damaged cell produces some novel molecule or some new molecular pattern that can act as an antigen; it cannot operate effectively against a damaged cell which merely ceases to synthesize some particular molecule unless this failure also results in some new and significant molecular pattern. Indeed, if the lymphocytes are efficient in destroying all antigenic mutant cells, then any tumour that manages to appear must either be non-antigenic or only weakly antigenic. It is this argument that leads to the pessimistic view of the possibility of cancer treatment based on immunological theory.

NON-MITOTIC TISSUES

It is obvious that no tumour can arise from any cells that are incapable of mitosis, and therefore one obvious advantage of a non-mitotic tissue is that it cannot contribute to the death of the animal through carcinogenesis. In such a tissue the accumulation of molecular damage may impose an increasing burden of malfunction, non-function, and even cell death, but it may be a long time before this reaches fatal proportions. Clearly in a long-lived animal the fewer the numbers of active or potentially active DNA or RNA molecules that remain in the cells the less is the chance that the random accumulation of damage will have serious consequences.

As Bullough (1965) has concluded: "in most tissues, for a variety of obvious reasons, it is necessary to replace cells rather than molecules, and the possibility of gene damage leading to cancer may be the price that such tissues must pay for the retention of a facultative genome capable of directing synthesis for mitosis".

THE CANCER PROBLEM

It must be obvious that so long as our knowledge of the mechanism of cellular homeostasis remains inadequate, no satisfactory explanation of the phenomenon of carcinogenesis is likely to be obtained. Carcinogenesis is first and foremost a biological and not a medical problem, and progress in cancer research has been severely hampered by the absence of any adequate biological framework which could be used as a guide to experimentation and into which new observations could be fitted. The fact that the present framework is so rudimentary "is largely the result of long neglect of the biological aspects of cancer" (Laws, 1966).

As described above, the problem of cellular homeostasis concerns the nature of the chemical messages that pass within and between cells and the manner in which the cells respond to these messages, while the problem of carcinogenesis concerns the way in which this cellular communication system may be disrupted. On a practical level this leads in turn to the two questions of whether in tumour cells the normal system of signals can be restored or strengthened, and whether when this is done these cells are likely to be able to respond.

The only two messenger molecules so far extracted and partly characterized are the epidermal and granulocytic chalones, and it has been shown that epidermal and granulocytic tumours are able to respond to them. The main importance of these substances lies in the fact that they are the first tissue-specific antimitotic agents to become available which are evidently natural components of the cellular homeostatic mechanism. So far only general cytotoxic substances have been used to inhibit cancer growth. As Roe and Ambrose (1966) have said, these substances "tend to be most effective against rapidly growing tumours" and they "have the drawback that they (also) damage those tissues of the body which are normally the most mitotically active". In consequence the cytotoxic drugs may do more damage to normal tissues than they do to the tumour. However, the practical value of the tissue chalones has yet to be proved.

To regard cancer solely from the viewpoint of a particular pattern of gene mutation, which expresses itself in increased growth only in certain particular environmental conditions, may seem a gross and inexcusable oversimplification of the problem. Certainly it is contrary to the usual views of the pathologist, who is impressed by the wide range in appearance and behaviour shown by the tumours he encounters. However, the single important characteristic of every tumour is its abnormal growth, and this indicates clearly that the primary damage is to the mechanism of cellular homeostasis, which is a mechanism that is evidently identical in all cells of all tissues.

In support of this it has been noted that in cancer-prone families the actual tumours that develop may be very diverse in their tissues of origin (Lynch et al., 1966). The inherited gene weaknesses evidently affect all tissue cells equally and it is a matter of chance where the extra mutations will occur that are needed for initiation.

In considering the practical problem of cancer the natural plan is to attack the long process of carcinogenesis in its weakest link, which is undoubtedly the period of dormancy. Probably it will never be possible to prevent initiation, and the destruction of a rapidly growing and metastasizing tumour may always present difficulties, but already it is possible to influence powerfully the duration of dormancy. In Tannenbaum's experiments (1947) with restricted diets the environment of the dormant cells was changed so dramatically that promotion did not occur within the normal life span of the mice. In time these may prove to be some of the most significant experiments in the whole history of cancer research. They show clearly that retarding agents may eliminate cancer, and an intensive search for suitable, and perhaps artificial, agents is now needed.

What is also needed is an intensive study of all aspects of cellular homeostasis, and in particular of the production of chemical messengers within cells and their transfer between cells (Iversen, 1965). It is the control mechanisms and the information systems within the tissue cells that contain "the nux of the problems both of cancer causation and treatment", and it is the study of these mechanisms that today forms "the most important growing point in cancer research" (Roe and Ambrose, 1966).

Summary

Anyone with a knowledge of any part of the vast literature on which this book is based will realize that the present argument represents a considerable oversimplification of the situation as it actually exists in all types of organisms from bacteria to mammals. It has, however, been the deliberate aim to under-emphasize the complexities, which are only too obvious, in an attempt to uncover those common themes that underlie all forms of cellular activity and therefore all forms of differentiation.

THE COMMON THEMES

Any new organism, whether created by binary fission or by the activation of an egg, inherits two essential structures. The first is the cell itself and the second is a set of coded instructions, mostly in the form of DNA molecules, whereby the cell is caused to function in a specific way. This DNA is mostly confined to the chromosomes, but in a eukaryote cell it is also found in such organelles as plastids and mitochondria.

THE ESSENCE OF LIFE

During the whole of evolution from the prokaryote cell onwards the basic biochemical pathways have remained essentially unchanged. As Sager (1965) has said: "Life began, I would speculate, with the emergence of a stabilized tripartite system: nucleic acids for replication, a photosynthetic or chemosynthetic system for energy conversion, and protein enzymes to catalyse the two processes". Sager has also suggested that in the earliest phases of life the only nucleic acid utilized may have been RNA, which may not even have been concentrated in any particular region of the cell. From time to time this system may have been refined, as for instance by the acquisition of the Krebs cycle and its ancillary processes when oxygen became a significant part of the atmosphere, but in essence it has remained remarkably unchanged. This extreme conservatism suggests that any radical modification would inevitably disrupt the whole organization of the cell.

It is this system that is the essence of life. It is the physical expression of this system that is visible as the cell, and it is the organization of the cell that must be inherited since it cannot be specified by the chromosomal DNA. This implies that the organization of the cell is self-coding, either with or without the assistance of extrachromosomal nucleic acids. Sonneborn (1963) has in fact suggested that at least part of the cellular coding system may be based on complex geometrical patterns of protein, carbohydrate, and lipid molecules. These patterns may contain an inherent quality of repetitiousness, which may be sufficient to provide the main driving force for cytoplasmic growth.

Unfortunately since the cellular coding systems are all so highly stable they are difficult to study. Any system that resists change also resists investigation.

THE FORM OF LIFE

To this basic cell mechanism has been added the DNA coding mechanism. Perhaps, as suggested by Sager (1965), the localization of most of the genetic information in one place in the form of chromosomal DNA may have been a relatively late refinement in the evolution of the cell. Today the DNA code appears to be a universal language, and the manner in which the code is read and acted upon appears to be identical in all types of cells. The evident uniformity of the DNA-RNA language and of its method of expression is dramatically illustrated by the claim, first, that messenger RNA from an insect when added to an extract of mammalian liver cells can direct the synthesis of a typical insect enzyme (Sekeris and Lang, 1964), and second, that messenger RNA from cow pituitary when added to an extract of bacteria can direct the synthesis of an adrenocorticotropic hormone (Todorov et al., 1966).

There are two aspects to gene activity: first, the genes specify the various metabolic enzymes, and second, they regulate their own activity according to the needs of the moment. Evidently this regulation is commonly, if not universally, achieved in terms of the effector-repressor-gene control mechanism that has been analysed by the microbiologists. It is this mechanism that underlies all forms of differentiation in micro-organisms.

From this it can be seen that all organisms, from bacteria onwards, consist of two parts: the essence, which is the biochemical and cellular system described above, and the form, which is the pattern of differentiation that stands as a homeostatic barrier between the essence and the environment. It is the form that is coded in the chromosomal genes and it is the form that has been the main subject of all evolutionary change from the bacteria to the highest plants and animals. This is recognized in the phrase *"omnis forma e DNA"*.

Indeed, the essence of evolution has been the selection of ever more effective forms within which the basic essence can be carried and protected and propagated. The most advanced organisms alive today are those with the most effective forms, that is with the most effective homeostatic mechanisms which confer the greatest versatility on their owners.

GENE CONTROL

However, it is gene control rather than gene expression that poses the most important biological problem of the moment. From a review of all the available evidence it is the main theme of this book that gene control is always achieved by the same kind of mechanism in all types of cells; genes are activated or inactivated in response to chemical messengers which travel within and between cells.

The situation is most clearly understood in the bacteria. Here the genetically-controlled phases of differentiation are transient and the chemical messengers commonly take the form of nutrients or metabolites. However, in the bacteria, as in the unicellular algae and the protozoans, it is probable that a number of chemical messengers are manufactured within the cells for the specific purpose of initiating and controlling the more complex genetic programmes that underlie such activities as cell division or sporulation. In the multicellular organisms most if not all of the chemical messengers which control genetic activity seem to be synthesized for similar specific purposes.

In a eukaryote cell gene activation always seems to depend on the formation of a zone of unwound DNA, the binding to this zone of RNA polymerase, the step-by-step advance of the polymerase as the unwound zone advances along the length of the operon, and the step-by-step synthesis and detachment of the messenger RNA molecule. Certainly the uncoiling of the DNA, accompanied by its separation from the chromosomal histone, is essential, and it can actually be seen both in lampbrush chromosomes and in chromosomal puffs.

It is this that has led to the suggestion that the histones may constitute the basic gene control mechanism of the eukaryotes, but there are good reasons for not accepting this and for suggesting that the primary function of the histones is to provide a skeletal support for the long and fragile DNA thread. It is probable that genes are in fact activated or inactivated by non-histone chemical messengers, as they undoubtedly are in the prokaryotes.

The main conclusion is that gene control in all its forms, whether leading to transient or to stable types of differentiation, is always achieved in a similar way in terms of the intracellular concentrations of a wide variety of chemical messengers, ranging from nutrient materials, through inducing substances and chalones, to hormones and pheromones.

THE CELL AS A WHOLE

Although it is generally accepted that the greater part of the hereditary information within a cell is chromosomal, as is clearly indicated by the weighty evidence of classical genetics, it is equally certain that the extrachromosomal hereditary information is also of great importance. In particular it has been stressed that genetic activity can only lead to a particular range of differentiation within the context of an appropriate cytoplasm. This cytoplasm may include specific molecules that are essential for a proper response to the genetic messages, or it may contain coded information that acts to modify these genetic messages, or it may even issue instructions of its own that are independent of the genes.

Certainly a cell is a functional unit, and following this argument to its extreme, Wright (1966) has even criticized the concept of a trigger mechanism in differentiation. She has stated that "in dealing with complicated phenomena which are brought about by varied and independent forces, searching for a single cause or trigger mechanism can only delay our eventual understanding of the problems involved". Differentiation must be considered as the expression of an interlocking system composed of essential genes, enzymes, and substrates, which must all have evolved together as part of one pattern. Wright concludes that in such a system it is improper to ask which part comes first, and that in considering induction "it is nearly impossible to say just where it begins".

These are important considerations, and clearly the interlocking role of the extrachromosomal cell constituents must never be discounted. At the moment, however, it appears that the argument has been overstated. If a cell is to function in a responsive way it must be able to adapt its state of differentiation to its environment. This means that it must be able to react to certain significant situations, the influence of which must break into the cell complex at some particular point if the genetic activity is to be modified. It is evidently at this point that the chemical messengers exert their influence, though whether they do so directly or indirectly is not always clear.

THE EVOLUTION OF DIFFERENTIATION

The oldest surviving cell type is that of the bacteria and the blue-green algae, both of which have fossil records extending back for more than 3000 million years. The evolution of the higher forms of life has depended, first, on the increasing complexity of the linkages within the groups of structural genes, and second, on the increasing number of regulator genes and the increasing complexity of the linkages between them. These changes have led, first, to a more complex type of cell, and later, to the multicellular organisms with their specialized tissues.

THE EVOLUTION OF THE CELL

Only two types of cell are known, the prokaryote and the eukaryote, and therefore the only known evolutionary advances in basic cell structure and organization are represented by the differences between them.

The most important single advance shown by the eukaryote cell is the great enlargement of the genome. In a prokaryote cell the repertoire of differentiation is severely restricted by the limited number of genes, and indeed the most complex form of differentiation developed by the bacteria may be sporulation, which may depend on the sequential activation of up to a hundred genes. The greatly enlarged genome of the eukaryote cell represents one of the greatest single advances in the whole course of evolution, since it permitted the organization of that infinitely wide range of differentiation which characterizes the higher plants and animals. One of these new forms of differentiation was the process of mitosis.

The second important advance in nuclear organization may have been the accidental outcome of a viral disease. The habit of conjugation in bacteria evidently began as a form of pathological differentiation dictated by a DNA virus to ensure the spread of the infection. However, with the incorporation of the viral genes into the "haploid" bacterial genome, a simple form of gene recombination was established, and this may ultimately have led to the habit of conjugation in the eukaryotes. If true, this may be the greatest single innovation bequeathed by the prokaryotes to their eukaryote successors. With the evolution of diploidy it opened the way for an efficient system of gene recombination, and with the evolution of multicellular organisms it opened the way for the sexual method of reproduction.

The third major evolutionary advance of the eukaryote cell involved the development of a variety of specific cell structures and organelles, many if not all of which may be self-coding. Such organelles as plastids and mitochondria, which use DNA coding systems and which show similarities with prokaryote cells, may also have originated as parasites or symbionts. Although all these types of self-coding systems possess a certain independence, it is clear that their activities always remain in step with those of the genes, and it follows that some system of communication must exist between them.

These advances in cellular structure, significant as they are, do not seem to have involved any significant modifications in the system of gene control by chemical messengers as found in the bacteria. However, with the evolution of the multicellular organisms, the system of chemical messengers was evidently greatly elaborated.

THE EVOLUTION OF TISSUES

The next major step forward was the evolution of systems of cooperation between cells to form a variety of multicellular organisms. This step must have been taken on many occasions, and the two most successful systems led to the higher plants and the higher animals.

In the metazoans the cells cooperated together to form tissues which in turn cooperated together to form a functional whole, and this involved two major innovations in the mechanism of gene control. First, while differentiation in unicellular organisms depends on chemical messengers that operate within the cell, differentiation in multicellular organisms depends on chemical messengers that pass from cell to cell. Second, while in unicellular organisms there is a strong selective advantage in labile forms of differentiation to meet the challenge of a changing environment, in multicellular organisms there is a strong selective advantage in stable forms of differentiation to ensure the continued existence of the tissues and organs.

Since a multicellular organism typically originates as a single cell, the production of a new individual depends on cell multiplication and on embryonic tissue differentiation. Such differentiation is the outcome of a complex pattern of gene activation and inactivation, which is dictated by the sequential responses of the genes to an array of specialized chemical messengers known as inducing agents. The general pattern is one of progressive and selective closure of what was originally a toti-potent genome.

After the stable differentiation of the tissue cells has been esta-blished, each cell commonly retains two alternative potentialities. It may differentiate for cell division or it may differentiate for ageing and for function (see Fig. 16). These are relatively labile forms of dif-ferentiation, and if ageing has not gone too far they are readily inter-changeable. The choice between them is evidently made in terms of the concentration of a tissue-specific antimitotic chalone, although it is possible that other relevant chemical messengers may also exist as in the case of the granulocytic antichalone described by Rytömaa and Kiviniemi (1967).

The potentiality of differentiation for mitosis is retained in any tissue in which cell damage may make cell replacement essential. In a long-lived animal this has its dangers, since a breakdown in the messenger-gene control mechanism may lead to cancer.

TISSUE ORGANIZATION

All typical animal tissues are in a constant state of flux, especially in terms of their mass and of their rate of function. Mass is maintained by the balance between the rate of cell production and the rate of cell loss; it is a function of the total body space; it is able to adjust itself to changing circumstances; and it is powerfully influenced by the genetic effects of the changing concentrations of chemical messengers, especially chalones, within the tissue and within the total body space. This is the mechanism of cellular homeostasis.

The rate of tissue function is determined partly by the supply of non-mitotic ageing cells, which automatically become functional, and partly by a mechanism which adjusts the rate of cell function to the demand. Together these constitute the mechanism of functional homeostasis.

Thus tissue homeostasis is a composite of cellular homeostasis and functional homeostasis.

From this there emerges yet another example of biological unifor-mity. All animal tissues are evidently constructed on exactly the same pattern, and this applies even to those, like striped muscle, which have lost their power of mitosis or those, like uterus, which have become hormone-dependent. All types of cells follow the same route round the mitotic cycle and along the ageing pathway to death (see Fig. 19), and it is only as they traverse the ageing pathway that their functional genes are fully activated. Tissues differ from each other only in two basic particulars: first, in the point of balance they maintain between the rate of cell production and the rate of cell loss (and this is strongly influenced by the chalone concentration), and second, in the type of functional genes that are activated.

The differences between tissues are differences of detail only; the similarities between tissues are profound. The basic pattern of tissue organization must have been established in the earliest metazoans, and the evolution of tissues has not involved any change in this basic pattern.

AGEING AND DEATH

One important aspect of this pattern of tissue organization is the genetically specified ageing and death of each cell, and the question arises whether this contributes in any way to the ageing and death of the whole organism. The death of the individual is the price a species must pay for its evolutionary progress, since only by the elimination of the old generation can a way be made for the selective testing of new mutations or of new gene combinations that have arisen in the new generation. For each species there must be an ideal life span which has been established by selection and which must therefore be gene-controlled.

The most obvious way to control genetically the life span of an individual is to utilize the existing genetic control of cellular ageing in one or more of the most important tissues. To be effective the mitotic genes must also be silenced to prevent any replacement of lost cells. Thus the life of an adult mammal may be determined by the time that is taken by the cells of the non-mitotic nervous and muscular tissues to traverse the length of the ageing pathways. As these cells begin to die there will be a steady reduction in tissue efficiency which must significantly diminish the competitive ability of the animal. It may have been primarily to achieve this end that the mammalian nerves and striped muscles have been deprived of their mitotic potential.

CHEMICAL MESSENGERS: THE NEW ENDOCRINOLOGY

The scientific study of "chemical messengers" began with the classical work of Bayliss and Starling (1902), although the existence of such substances was suspected long before. The chemical messenger they described was secretin "which is absorbed from the (duodenal) cells by the blood-current, and is carried to the pancreas, where it acts as a specific stimulus to the pancreatic cells, exciting a secretion of pancreatic juice proportional to the amount of secretin present". Later, Starling (1906) proposed that such chemical messengers should be called hormones: "In Anbetracht der ausgesprochen charakteristischen Eigenschaften dieser Körpergruppe und der wichtigen Aufgaben die derselben im Organismus der höheren Tiere zufallen, schlage ich vor, diesen Substanzen einen eigenen Namen zu geben, und ich werde sie deshalb fernerhin in diesem Vortrage als Hormone (von ὁρμάω =ich reize oder rege an) bezeichnen".

Still later Schäfer (1916) was critical of the term "hormone" since it had led Starling to the absurdity of distinguishing between "erregende Hormone" and "hemmende Hormone", that is between "excitants which excite" and "excitants which prevent excitation". He regretted that the original term "chemical messenger" had not been allowed to develop into the term "hermone" (from 'Ερμῆς meaning the messenger god Hermes), which would have avoided the difficulty of having to distinguish between stimulants and inhibitors.

However, the word hormone has now been universally accepted to describe that type of chemical messenger which is "normally produced in the cells of some part of the body, and carried by the blood-stream to distant parts, which it affects for the good of the organism as a whole" (Starling, 1914). The study of such substances is the concern of endocrinology today. However, it is now becoming obvious that hormones cannot logically be separated, except for convenience, from all those other chemical messengers which are "secreted internally" within the cells and tissues of unicellular and multicellular organisms. It is now necessary to consider all the chemical messengers that circulate within and between the cells of all known organisms as constituting in each case a single, complex, homeostatic system.

The foundations of the system appear to be those revealed by the microbiologists. In the bacteria the structural genes are evidently activated or inactivated by specific repressor molecules which are specified by regulatory genes. These repressor molecules may be regarded as the basic chemical messengers, and it seems likely that no prokaryote or eukaryote cell could function without them.

Little is known of these basic chemical messengers, although one, the repressor of the *lac* operon of *E. coli*, has been partly isolated and appears to be a protein (Gilbert and Müller-Hill, 1966). Indeed, it is a tenet of the Jacob-Monod theory that repressors must be proteins which possess at least two active sites. It is by binding to one of these sites that an effector molecule is thought to produce an allosteric change in the shape of the repressor molecule and thus to alter the shape of the second active site. It is the function of this second site to combine with the operator gene, and the deformation that it suffers in the presence of the effector either permits or prevents this action. Although this theory has not yet been tested with any repressor molecule, the existence of two such active sites has been shown in the molecule of aspartate transcarbamidase, the one having an affinity for the substrate and the other for an inhibitor (Gerhart, 1964).

Thus the effector substances form a second tier in the hierarchy of the chemical messengers. They are evidently of two kinds: the first is a simple foodstuff or metabolite, as has been shown particularly clearly in the bacteria, while the second is a molecule that is specially synthesized by the cell to activate or inactivate a particular repressor molecule. In the unicellular organisms both kinds of effector exist; in the multicellular organisms the emphasis is on specially synthesized effectors, which must clearly be regarded as part of the endocrine system.

The common feature of effector substances is that they activate or inactivate one particular part of the genome. Those effectors which are specially synthesized are mostly chemically unknown, but they evidently range from the initiators and inhibitors of cell division, of sporulation, and of conjugation in the unicellular organisms to the initiators and inhibitors of tissue formation and the controllers of tissue homeostasis in the multicellular organisms. In the metazoans they appear to include both the embryonic inducers and the tissue chalones.

The third tier in the hierarchy of chemical messengers is formed by the hormones. They have clearly been added mainly to enable certain tissues to respond to events that originate outside the animal, although inducing agents may sometimes be utilized for this purpose instead. The involvement of inducing agents is best illustrated by the poietins, which increase the supply of erythrocytes in oxygen lack, or the supply of granulocytes during infection, by inducing the final step in tissue differentiation in a stem cell population.

However, in the higher animals the hormone type of response to externally originating stimuli is more common. This response may be relatively simple, as when the stomach wall secretes the hormone gastrin in response to food entering the stomach, or when the islets of Langerhans secrete the hormone insulin in response to a rise in the blood sugar level. However, in many hormone systems, the message from the environment enters through the main sense organs and passes via the nervous system. The simplest such case may be when a frightening situation causes the secretion of adrenalin; the most complex may be when increasing day-length acts via the eyes and the brain, via the neurohormones of the hypothalamus, and via the pituitary and its gonadotropins to stimulate the growth of the gonads, which then secrete their own hormones to induce the growth of the accessory sexual organs and the secondary sexual structures.

Thus the hormone mechanisms range from a simple homeostatic response to food to a complex nerve-hormone mechanism, which by controlling the breeding season forms part of the homeostatic control of the population.

The manner of hormone action is not yet known and may not always be the same. Many hormones, however, achieve their effects by altering gene activity, although in these cases it seems that they operate through some mechanism that stands between them and the genes. This mechanism may consist of, or at least include, the chalone mechanisms of the target tissues.

Merging with the hormones, and sometimes perhaps forming a fourth level in the hierarchy of the chemical messengers, are the pheromones. These operate to maintain a homeostatic balance within societies of animals. Their manner of action is unknown, but certainly in insects such as bees and termites they induce changes at the tissue level.

In any species it is evident that the hormone and pheromone systems are the most recently evolved parts of the system of chemical messengers, and it is evident, too, that the hormone molecules themselves have not been subject to much evolutionary change (Barrington, 1964). Hormone systems appear to depend primarily on the evolution of appropriate tissue responses. It is these responses that enable the species to react adequately to those environmental signals which in the course of its evolutionary history it has found to be of special significance.

It is therefore the task of the new endocrinology to consider the nature and function of a hierarchy of chemical messengers, and not just those of the hormones alone. All chemical messengers act as links in homeo-

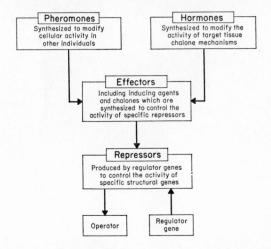

FIG. 32. The new endocrinology. The internally secreted chemical messengers which are now becoming known appear to operate at a number of different levels and so to form a hierarchy of control mechanisms.

static control mechanisms, and as evolution has proceeded and homeostasis has become more complex and more efficient, new levels of
control have been added (see Fig. 32). It now begins to appear that most,
if not all, of the chemical messengers may act, directly or indirectly, to
modify gene action and therefore to modify the state of differentiation
of cells and of tissues. Indeed, the whole of cellular evolution has been
concerned primarily with the ever increasing sophistication of gene
control systems, which has necessitated an ever increasing number and
variety of chemical messengers.

However, the most important single conclusion must be that this
evolution of cellular differentiation, dramatic as its consequences have
been, has involved nothing more than an increasing elaboration of an
ancient theme of gene control that may have originated in some
primitive prokaryote cell.

References

Abercrombie, M. (1957). Localized formation of new tissue in an adult mammal. *Symp. Soc. Exp. Biol.* **11**, 235–254.

Abercrombie, M. (1965). Cellular interactions in development. *In* "Ideas in Modern Biology". (Ed. Moore, J. A.) Doubleday, New York.

Agrell, I. (1964). Physiological and biochemical changes during insect development. *In* "The Physiology of Insecta", Vol. 1. (Ed. Rockstein, M.) Academic Press, New York.

Alexander, P. (1966). The role of tumour-specific antigens in the genesis, development and control of malignant disease. *In* "The Biology of Cancer". (Eds Ambrose, E. J. and Roe, F. J. C.) Van Nostrand, London.

Alho, A. (1961). Regeneration capacity of the submandibular gland in rat and mouse. *Acta Pathol. Microbiol. Scand. Suppl.* **149**, 1–84.

Allfrey, V. G., Littan, V. C. and Mirsky, A. E. (1963). On the role of histones in regulating ribonucleic acid synthesis in the cell nucleus. *Proc. Nat. Acad. Sci. U.S.* **49**, 414–420.

Allsopp, A. (1964). Shoot morphogenesis. *Ann. Rev. Plant Physiol.* **15**, 225–254.

Ambrose, E. J. and Roe, F. J. C. (eds) (1966). "The Biology of Cancer". Van Nostrand, London.

Ames, B. N. and Hartman, P. E. (1963). The histidine operon. *Cold Spring Harbor Symp. Quant. Biol.* **28**, 349–356.

Anderson, M. R. (1964). Variations in the rate of induction of chemical carcinogenesis according to differing psychological states in rats. *Nature* **204**, 55–56.

Argyris, T. S. (1966). Enzyme induction and the control of growth. Personal communication.

Argyris, T. S. and Trimble, M. E. (1964). The growth promoting effects of damage in the damaged and contralateral kidneys of the mouse. *Anat. Record* **150**, 1–10.

Bantock, C. (1961). Chromosome elimination in Cecidomyidae. *Nature* **190**, 466–467.

Barghoorn, E. S. and Schopf, J. W. (1966). Microorganisms three billion years old from the Precambrian of South Africa. *Science* **152**, 758–762.

Barghoorn, E. S. and Tyler, S. A. (1965). Microorganisms from the Gunflint Chert. *Science* **147**, 563–577.

Barrington, E. J. W. (1964). "Hormones and Evolution". English University Press, London.

Barth, L. G. (1941). Neural differentiation without organizer. *J. Exp. Zool.* **87**, 371–383.

Bayliss, W. M. and Starling, E. H. (1902). The mechanism of pancreatic secretion. *J. Physiol. (London)* **28**, 325–353.

Beale, G. H. (1954). "The Genetics of *Paramecium aurelia*". Cambridge University Press.

Beale, G. H. (1964). Genes and cytoplasmic particles in *Paramecium*. *In* "Cellular Control Mechanisms and Cancer". (Eds Emmelot, P. and Mühlock, O.) Elsevier, Amsterdam.

Beermann, W. (1952). Chromomerenkonstanz und spezifische Modifikationen der Chromosomenstruktur in der Entwicklung und Organdifferenzierung von *Chironomus tentans*. *Chromosoma* 5, 139–198.

Beermann, W. (1952). Chromosomenstruktur und Zelldifferenzierung in der Speicheldrüse von *Trichocladius vitripennis*. *Z. Naturforsch.* 7b, 237–242.

Beermann, W. and Clever, U. (1964). Chromosome puffs. *Sci. Am.* 210, 50–58.

Beltchev, B. G. and Tsanev, R. (1966). Ribonucleic acid degrading activity of rat liver microsomes following partial hepatectomy. *Nature* 212, 531–532.

Berenblum, I. (1941). The mechanism of carcinogenesis. *Cancer Res.* 1, 807–814.

Berenblum, I. (1954). The probable nature of promoting action and its significance in the understanding of the mechanism of carcinogenesis. *Cancer Res.* 14, 471–477.

Berenblum, I. and Shubik, P. (1947). A new quantitative approach to the study of the stages of chemical carcinogenesis in mouse's skin. *Brit. J. Cancer* 1, 383–391.

Berliner, D. L. (1964). Biotransformation of corticosteroids as related to inflammation. *Ann. N. Y. Acad. Sci.* 116, 1078–1083.

Bern, H. A. (1966). On the production of hormones by neurones and the role of neurosecretion in neuroendocrine mechanisms. *Symp. Soc. Exp. Biol.* 20, 325–345.

Bernal, J. D. (1951). "The Physical Basis of Life". Routledge and Kegan Paul, London.

Berrill, N. J. (1950). "The Tunicata". Ray Society, London.

Bertalanffy, F. D. and Lau, C. (1962). Cell renewal. *Intern. Rev. Cytol.* 13, 357–366.

Betz, E. H., Firket, H. and Reznik, M. (1966). Some aspects of muscle regeneration. *Intern. Rev. Cytol.* 19, 203–226.

Billingham, R. E. (1958). A reconsideration of the phenomenon of hair neogenesis with particular reference to the healing of cutaneous wounds in adult mammals. *In* "The Biology of Hair Growth". (Eds Montagna, W. and Ellis, R. A.) Academic Press, London.

Bishop, J., Favelukes, G., Schweet, R. and Russel, E. (1961). Control of specificity in haemoglobin synthesis. *Nature* 191, 1365–1368.

Blecher, M. and White, A. (1958). Effects of various steroids and metabolic inhibitors on the incorporation of glycine-2-C^{14} into total proteins and nucleic acids of normal and malignant lymphocytes *in vitro*. *J. Biol. Chem.* 233, 1162–1168.

Blumenthal, H. T. (1950). The nature of cycle variations in mitotic activity. *Growth* 14, 231–249.

Boggs, D. R. (1966). Homeostatic regulatory mechanisms of hematopoiesis. *Ann. Rev. Physiol.* 28, 39–56.

Bonner, J. T. (1963). Epigenetic development in the cellular slime moulds. *Symp. Soc. Exp. Biol.* 17, 341–358.

Bonner, J. T. (1965). "The Molecular Biology of Development". Clarendon Press, Oxford.

Bonner, J. T. and Dodd, M. R. (1962). Aggregation territories in the cellular slime moulds. *Biol. Bull.* 122, 13–24.

Bonner, J. T. and Ts'o, P. O. P. (eds) (1964). "The Nucleohistones". Holden-Day, San Francisco.

Bonner, J. T., Kelso, A. P. & Gillmor, R. G. (1966). A new approach to the problem of aggregation in the cellular slime moulds. *Biol. Bull.* **130**, 28–42.

Boutwell, R. K., Brush, M. K. and Rusch, H. P. (1948). Some physiological effects associated with chronic caloric restriction. *Am. J. Physiol.* **154**, 517–524.

Boveri, T. (1899). "Die Entwicklung von *Ascaris megalocephala* mit besonderer Rücksicht auf die Kernverhältnisse". Gustav Fischer, Jena.

Brachet, J. (1960). "The Biochemistry of Development". Pergamon Press, London.

Brachet, J. (1963). The role of the nucleic acids in the processes of induction, regulation and differentiation in the amphibian embryo and the unicellular alga, *Acetabularia mediterranea*. *In* "Biological Organization at the Cellular and Supercellular Level". (Ed. Harris, R. J. C.) Academic Press, London.

Brachet, J. (1967). Protein synthesis in the absence of the nucleus. *Nature* **213**, 650–655.

Brandes, D., Bertini, F. and Smith, E. W. (1965). Role of lysosomes in cellular lytic processes. *Exp. Mol. Path.* **4**, 245–265.

Breedis, C. (1954). Regeneration of hair follicles and sebaceous glands from the epithelium of scars in the rabbit. *Cancer Res.* **14**, 575–579.

Brenner, S. (1965). Theories of gene regulation. *Brit. Med. Bull.* **21**, 244–248.

Briggs, R. and King, T. J. (1952). Transplantation of living nuclei from blastula cells into enucleated frogs' eggs. *Proc. Nat. Acad. Sci. U. S.* **38**, 455–463.

Briggs, R. and King, T. J. (1957). Changes in the nuclei of differentiating endoderm cells as revealed by nuclear transplantation. *J. Morphol.* **100**, 269–312.

Briggs, R. and King, T. J. (1959). Nucleocytoplasmic interactions in eggs and embryos. *In* "The Cell", Vol. 1. (Eds Brachet, J. and Mirsky, A. E.) Academic Press, New York.

Briggs, R. and King, T. J. (1960). Nuclear transplantation studies on the early gastrula (*Rana pipiens*). I. Nuclei of presumptive endoderm. *Develop. Biol.* **2**, 252–270.

Brodsky, V. Y., Khruschov, N. G. and Kusch, A. A. (1964). Non-uniformity of premitotic DNA reduplication in the cells of mammals. *Bull. Exp. Biol. Med. (USSR) (English Transl.)* **3**, 94–97.

Brody, H. (1955). Organization of the cerebral cortex. A study of aging in the human cerebral cortex. *J. Comp. Neurol.* **102**, 511–556.

Brown, R. (1963). Cellular differentiation in the root. *Symp. Soc. Exp. Biol.* **17**, 1–17.

Brown, W. M. C. and Tough, I. M. (1963). Cytogenetic studies in chronic myeloid leukaemia. *Advan. Cancer Res.* **7**, 351–381.

Bucher, N. L. R. (1963). Regeneration of mammalian liver. *Intern. Rev. Cytol.* **15**, 245–300.

Bucher, N. L. R. and Swaffield, M. N. (1964) The rate of incorporation of labeled thymidine into the deoxyribonucleic acid of regenerating rat liver in relation to the amount of liver excised. *Cancer Res.* **24**, 1611–1625.

Bucher, N. L. R., Scott, J. F. and Aub, J. C. (1951). Regeneration of the liver in parabiotic rats. *Cancer Res.* **11**, 457–465.

Bullough, W. S. (1946). Mitotic activity in the adult female mouse. *Phil. Trans. Roy. Soc. London, Ser. B* **231**, 453–516.

Bullough, W. S. (1947). Hermaphroditism in the lower vertebrates. *Nature* **160**, 9–11.

Bullough, W. S. (1950). Mitotic activity in the tissues of dead mice, and in tissues kept in physiological salt solutions. *Exp. Cell Res.* **1**, 410–420.

Bullough, W. S. (1961). "Vertebrate Reproductive Cycles". (2nd Edition.) Methuen, London.

Bullough, W. S. (1962). The control of mitotic activity in adult mammalian tissues. *Biol. Rev. Cambridge Phil. Soc.* **37**, 307–342.

Bullough, W. S. (1963). Analysis of the life-cycle in mammalian cells. *Nature* **199**, 859–862.

Bullough, W. S. (1964). Growth regulation by tissue-specific factors, or chalones. *In* "Cellular Control Mechanisms and Cancer". (Eds Emmelot, P. and Mühlbock, O.) Elsevier, Amsterdam.

Bullough, W. S. (1965). Mitotic and functional homeostasis. *Cancer Res.* **25**, 1683–1727.

Bullough, W. S. and Ebling, F. J. (1952). Cell replacement in the epidermis and sebaceous glands of the mouse. *J. Anat.* **86**, 29–34.

Bullough, W. S. and Eisa, E. A. (1950). The effects of a graded series of restricted diets on epidermal mitotic activity in the mouse. *Brit. J. Cancer* **4**, 321–328.

Bullough, W. S. and Laurence, E. B. (1958). The mitotic activity of the follicle. *In* "The Biology of Hair Growth". (Eds Montagna, W. and Ellis, R. A.) Academic Press, London.

Bullough, W. S. and Laurence, E. B. (1960). The control of epidermal mitotic activity in the mouse. *Proc. Roy. Soc. (London), Ser. B* **151**, 517–536.

Bullough, W. S. and Laurence, E. B. (1960). The control of mitotic activity in mouse skin. Dermis and hypodermis. *Exp. Cell Res.* **21**, 394–405.

Bullough, W. S. and Laurence, E. B. (1961). Stress and adrenaline in relation to the diurnal cycle of epidermal mitotic activity in adult male mice. *Proc. Roy. Soc. (London), Ser. B* **154**, 540–556.

Bullough, W. S. and Laurence, E. B. (1964). The production of epidermal cells. *Symp. Zool. Soc. Lond.* **12**, 1–23.

Bullough, W. S. and Laurence, E. B. (1964). Duration of epidermal mitosis *in vitro. Exp. Cell Res.* **35**, 629–641.

Bullough, W. S. and Laurence, E. B. (1964). Mitotic control by internal secretion. The role of the chalone-adrenalin complex. *Exp. Cell Res.* **33**, 176–194.

Bullough, W. S. and Laurence, E. B. (1966). The diurnal cycle in epidermal mitotic duration and its relation to chalone and adrenalin. *Exp. Cell Res.* **43**, 343–350.

Bullough, W. S. and Laurence, E. B. (1966). Tissue homeostasis in adult mammals. *In* "Advances in Biology of Skin", Vol. 7. Carcinogenesis. (Eds Montagna, W. and Dobson, R. L.) Pergamon Press, New York.

Bullough, W. S. and Laurence, E. B. (1967). Epigenetic mitotic control. *In* "Control of Cellular Growth in Adult Organisms". (Eds Teir, H. and Rytömaa, T.) Academic Press, London.

Bullough, W. S. and Oordt, G. J. van. (1950). The mitogenic actions of testosterone propionate and of oestrone on the epidermis of the adult male mouse. *Acta Endocrinol.* **4**, 291–305.

Bullough, W. S. and Rytömaa, T. (1965). Mitotic homeostasis. *Nature* **205**, 573–578.

Bullough, W. S., Hewett, C. L. and Laurence, E. B. (1964). The epidermal chalone: a preliminary attempt at isolation. *Exp. Cell Res.* **36**, 192–200.

Bullough, W. S., Laurence, E. B., Iversen, O. H. and Elgjo, K., (1967). The vertebrate epidermal chalone. *Nature* **214**, 578–580.

Burch, P. R. J. (1962). A biological principle and its converse: some implications for carcinogenesis. *Nature* **195**, 241–243.

Burch, P. R. J. (1963). Human cancer: Mendelian inheritance or vertical transmission. *Nature* **197**, 1042–1045.

Burnet, F. M. (1957). Cancer—a biological approach. *Brit. Med. J.* **1**, 779–786.

Burnet, F. M. (1959). "The Clonal Selection Theory of Acquired Immunity". Cambridge University Press.

Burnet, F. M. (1962). "The Integrity of the Body". Oxford University Press.

Burnet, F. M. (1964). Immunological factors in the process of carcinogenesis. *Brit. Med. Bull.* **20**, 154–158.

Burnet, F. M. (1966). A possible genetic basis for specific pattern in antibody. *Nature* **210**, 1308–1310.

Busch, H., Steele, W. J., Hnilica, L. S., Taylor, C. W. and Mavioglu, H. (1963). Biochemistry of histones and the cell cycle. *J. Cellular Comp. Physiol.* **62**, suppl. 1, 95–110.

Butler, J. A. V. (1965). Role of histones and other proteins in gene control. *Nature* **207**, 1041–1042.

Caffery, J. M., Whichard, L. and Irvin, J. L. (1964). Effect of histones on the induction of two liver enzymes by hydrocortisone. *Arch. Biochem. Biophys.* **108**, 364–365.

Cahn, R. D. and Cahn, M. B. (1966). Hereditability of cellular differentiation: clonal growth and expression of differentiation in retinal pigment cells *in vitro*. *Proc. Nat. Acad. Sci. U. S.* **55**, 106–114.

Cairns, J. (1963). The bacterial chromosome and its manner of replication as seen by autoradiography. *J. Mol. Biol.* **6**, 208–213.

Calkins, G. N. (1933). "The Biology of the Protozoa". Baillière, Tindall and Cox, London.

Cattaneo, S. M., Quastler, H. and Sherman, F. G. (1961). Proliferative cycle in the growing hair follicle of the mouse. *Nature* **190**, 923–924.

Changeux, J. P. (1965). The control of biochemical reactions. *Sci. Am.* **212**, 36–45.

Chase, H. B. (1954). Growth of the hair. *Physiol. Rev.* **34**, 113–126.

Chase, H. B. and Eaton, G. J. (1959). The growth of hair follicles in waves. *Ann. N. Y. Acad. Sci.* **83**, 365–368.

Clever, U. (1964). Puffing in giant chromosomes of Diptera and the mechanism of its control. *In* "The Nucleohistones". (Eds Bonner, J. and Ts'o, P.) Holden-Day, San Francisco.

Clever, U. (1964). Actinomycin and puromycin: effects on sequential gene activation by ecdysone. *Science* **146**, 794–795.

Clever, U. (1965). The effect of ecdysone on gene activity patterns in giant chromosomes. *In* "Mechanisms of Hormone Action". (Ed. Karlson, P.) Academic Press, London.

Clever, U. (1966). Gene activity patterns and cellular differentiation. *Am. Zool.* **6**, 33–42.

Clever, U. and Karlson, P. (1960). Induktion von Puff-veränderungen in den Speicheldrüsenchromosomen von *Chironomus tentans* durch Ecdyson. *Exp. Cell Res.* **20**, 623–626.

Comfort, A. (1964). "Ageing, the Biology of Senescence". Routledge and Kegan Paul, London.

Conney, A. H. and Gilman, A. G. (1963). Puromycin inhibition of enzyme induction by 3-methylcholanthrene and phenobarbital. *J. Biol. Chem.* **238**, 3682–3685.

Corliss, J. O. (1961). "The Ciliated Protozoa". Pergamon Press, London.

G

Coulombre, A. J. (1965). The eye. *In* "Organogenesis". (Eds DeHaan, R. L. and Ursprung, H.) Holt, Rinehart and Winston, New York.

Craddock, C. G. (1960). Production and distribution of granulocytes and the control of granulocyte release. *In* "Ciba Foundation Symposium on Haemopoiesis". (Eds Wolstenholme, G. E. W. and O'Connor, M.) Churchill, London.

Crump, L. M. (1950). The influence of the bacterial environment in the excystment of amoebae from soil. *J. Gen. Microbiol.* 4, 16–21.

Curtis, A. S. G. (1960). Cortical grafting in *Xenopus laevis. J. Embryol. Exp. Morphol* 8, 163–173.

Curtis, A. S. G. (1962). Morphogenetic interactions before gastrulation in the amphibian, *Xenopus laevis*—the cortical field. *J. Embryol. Exp. Morphol.* 10, 410–422.

Curtis, H. J. (1963). Biological mechanisms underlying the aging processes. *Science* 141, 686–694.

Curtis, H. J. (1966). "Biological Mechanisms of Aging". C. C. Thomas, Springfield, Illinois.

Dalcq, A. and Pasteels, J. (1937). Une conception nouvelle des bases physiologiques de la morphogénèse. *Arch. Biol. (Liège)* 48, 669–710.

Daniel, J. C. and Olson, J. D. (1966). Cell movement, proliferation and death in the formation of the embryonic axis of the rabbit. *Anat. Record* 156, 123–127.

Davidson, E. H. (1964). Differentiation in monolayer tissue culture cells. *Advan. Genet.* 12, 143–280.

Davidson, E. H. (1965). Hormones and genes. *Sci. Am.* 212, 36–45.

Davidson, E. H., Allfrey, V. G. and Mirsky, A. E. (1963). Gene expression in differentiated cells. *Proc. Nat. Acad. Sci. U. S.* 49, 53–60.

Dean, C. J., Feldschreiber, P. and Lett, J. T. (1966). Repair of x-ray damage to the DNA in *Micrococcus radiodurans. Nature* 209, 49–52.

Dee, J. (1962). Recombination in a myxomycete, *Physarum polycephalum* Schw. *Genet. Res.* 3, 11–23.

Defendi, V. and Manson, L. A. (1963). Analysis of the life-cycle in mammalian cells. *Nature* 198, 359–361.

Demerec, M. and Hartman, P. E. (1959). Complex loci in microorganisms. *Ann. Rev. Microbiol.* 13, 377–406.

Deuchar, E. M. (1966). "Biochemical Aspects of Amphibian Development'. Methuen, London.

DuPraw, E. J. (1966). Evidence for a "folded-fibre" organization in human chromosomes. *Nature* 209, 577–581.

DuPraw, E. J. and Rae, P. M. M. (1966). Polytene chromosome structure in relation to the "folded-fibre" concept. *Nature* 212, 598–600.

Durward, A. and Rudall, K. M. (1949). Studies on hair growth in the rat. *J. Anat.* 83, 325–335.

Durward, A. and Rudall, K. M. (1958). The vascularity and patterns of growth of hair follicles. *In* "The Biology of Hair Growth". (Eds Montagna, W. and Ellis, R. A.) Academic Press, London.

Duve, C. de (1963). The lysosome concept. *In* "Lysosomes". (Eds Reuck, A. V. S. de and Cameron, M. P.) Churchill, London.

Duve, C. de (1963). The lysosome. *Sci. Am.* 208, 64–72.

Ebling, F. J. (1957). The action of testosterone and oestradiol on the sebaceous glands and epidermis of the rat. *J. Embryol. Exp. Morph.* 5, 74–82.

Ebling, F. J. (1963). Hormonal control of sebaceous glands in experimental animals. *In* "Advances in Biology of Skin", Vol. 4. (Eds Montagna, W., Ellis, R. A. and Silver, A. F.) Pergamon Press, New York.

Ebling, F. J. and Johnson, E. (1961). Systemic influence on activity of hair follicles in skin homografts. *J. Embryol. Exp. Morph.* **9**, 285–293.

Ebling, F. J. and Johnson, E. (1964). The control of hair growth. *Symp. Zool. Soc. Lond.* **12**, 97–126.

Echlin, P. (1966). Origins of photosynthesis. *Sci. J.* **2**, 142–147.

Echlin, P. and Morris, I. (1965). The relationship between blue-green algae and bacteria. *Biol. Rev. Cambridge Phil. Soc.* **40**, 143–187.

Edelman, I. S., Bogoroch, R. and Porter, G. A. (1963). On the mechanism of action of aldosterone on sodium transport: the role of protein synthesis. *Proc. Nat. Acad. Sci. U. S.* **50**, 1169–1177.

Edgar, R. S. and Epstein, R. H. (1965). The genetics of a bacterial virus. *Sci. Am.* **212**, 70–78.

Edström, J. E. and Beermann, W. (1962). The base composition of nucleic acids in chromosome puffs, nucleoli, and cytoplasm of *Chironomus* salivary gland cells. *J. Cell Biol.* **14**, 371–379.

El-Antably, H. M. M. and Wareing, P. F. (1966). Stimulation of flowering in certain short-day plants by abscisin. *Nature* **210**, 328–329.

Epifanova, O. I. (1966). Mitotic cycles in estrogen-treated mice: a radioautographic study. *Exp. Cell Res.* **42**, 562–577.

Erickson, R. O. (1959). Patterns of cell growth and differentiation in plants. *In* "The Cell", Vol. 1. (Eds Brachet, J. and Mirsky, A. E.) Academic Press, New York.

Fincham, J. R. S. and Day, P. R. (1963). "Fungal Genetics". Blackwell, Oxford.

Finegold, M. J. (1965). Control of cell multiplication in epidermis. *Proc. Soc. Exp. Biol. Med.* **119**, 96–100.

Fischberg, M. and Blackler, A. W. (1961). How cells specialize. *Sci. Am.* **205**, 124–133.

Fischberg, M. and Blackler, A. W. (1963). Loss of nuclear potentiality in the soma versus preservation of nuclear potentiality in the germ line. *In* "Biological Organization at the Cellular and Supercellular level". (Ed. Harris, R. J. C.) Academic Press, London.

Fischberg, M. and Blackler, A. W. (1963). Nuclear changes during the differentiation of animal cells. *Symp. Soc. Exp. Biol.* **17**, 138–156.

Fortier, C. (1963). Hypothalamic control of anterior pituitary. *In* "Comparative Endocrinology", Vol. 1. (Eds Euler, U. S. von and Heller, H.) Academic Press, London.

Foulds, L. (1963). Some problems of differentiation and integration in neoplasia. *In* "Biological Organization at the Cellular and Supercellular Level". (Ed. Harris, R. J. C.) Academic Press, London.

Foulds, L. (1964). Tumour progression and neoplastic development. *In* "Cellular Control Mechanisms and Cancer". (Eds Emmelot, P. and Mühlbock, O.) Elsevier, Amsterdam.

Foulds, L. (1965). Multiple aetiological factors in neoplastic development. *Cancer Res.* **25**, 1339–1347.

Fox, S. W. (ed.) (1965). "The Origins of Prebiological Systems". Academic Press, New York.

Frenster, J. H. (1965). Nuclear polyanions as de-repressors of synthesis of ribonucleic acid. *Nature* **206**, 680–683.

Frenster, J. H. (1965). A model of specific de-repression within interphase chromatin. *Nature* **206**, 1269–1270.

Fuller, R. H., Brown, E. and Mills, C. A. (1941). Environmental temperatures and spontaneous tumors in mice. *Cancer Res.* **1**, 130–133.

Gabourel, J. D. and Aronov, L. (1962). Growth inhibitory effects of hydrocortisone on mouse lymphoma ML-388 *in vitro*. *J. Pharmacol. Exp. Therap.* **136**, 213–221.

Galicich, J. H., Halberg, F. and French, L. A. (1963). A circadian adrenal cycle in C mice kept without food and water for a day and a half. *Nature* **197**, 811–813.

Gall, J. G. (1963). Chromosomes and cytodifferentiation. *In* "Cytodifferentiation and Macromolecular Synthesis". (Ed. Locke, M.) Academic Press, New York.

Garen, A. and Echols, H. (1962). Genetic control of induction of alkaline phosphatase synthesis in *E. coli*. *Proc. Nat. Acad. Sci. U. S.* **48**, 1398–1402.

Gelboin, H. V. and Blackburn, N. R. (1963). The stimulatory effect of 3-methylcholanthrene on microsomal amino acid incorporation and benzpyrene hydroxylase activity and its inhibition by actinomycin D. *Biochim. Biophys. Acta* **72**, 657–660.

Gerhart, J. C. (1964). Subunits for control and catalysis in aspartate transcarbamylase. *Brookhaven Symp. Biol.* **17**, 222–231.

Gibbs, H. F. (1941). A study of the post-natal development of the skin and hair of the mouse. *Anat. Record* **80**, 61–79.

Gilbert, L. I. (1964). Physiology of growth and development: endocrine aspects. *In* "The Physiology of Insecta", Vol. 1. (Ed. Rockstein, M.) Academic. Press, New York.

Gilbert, W. and Müller-Hill, B. (1966). Isolation of the *lac* repressor. *Proc. Nat. Acad. Sci. U. S.* **56**, 1891–1898.

Gillman, T., Hathorn, M. and Penn, J. (1956). Actions of cortisone on cutaneous application of methylcholanthrene. *Brit. J. Cancer* **10**, 394–400.

Glaessner, M. F. (1962). Pre-Cambrian fossils. *Biol. Rev. Cambridge Phil. Soc.* **37**, 467–494.

Glinos, A. D. (1960). Environmental feedback control of cell division. *Ann. N. Y. Acad. Sci.* **90**, 592–602.

Glinos, A. D., Bucher, N. L. R. and Aub, J. C. (1951). The effect of regeneration on tumor formation in rats fed 4-dimethyl-aminoazobenzene. *J. Exp. Med.* **93**, 313–324.

Glucksmann, A. (1951). Cell deaths in normal vertebrate ontogeny. *Biol. Rev. Cambridge Phil. Soc.* **26**, 59–86.

Godefroy, F. (1964). Influence du jeûne sur la teneur en adrénaline et en noradrénaline des surrénales et de l'urine. *Compt. Rend. Soc. Biol.* **158**, 693–696.

Goldstein, L. (1965). Actinomycin D inhibition of the adaptation of renal glutamine-deaminating enzymes in the rat. *Nature* **205**, 1330–1331.

Goldstein, L., Stella, E. J. and Knox, W. E. (1962). The effect of hydrocortisone on tyrosine-α-ketoglutarate transaminase and tryptophan pyrrolase activities in the isolated, perfused rat liver. *J. Biol. Chem.* **237**, 1723–1726.

Goodman, G. J. (1957). Effects of one tumor upon the growth of another. *Proc. Am. Ass. Cancer Res.* **2**, 207.

Goodwin, T. W. (1964). The plastid pigments of flagellates. *In* "Biochemistry and Physiology of Protozoa". (Ed. Hutner, S. H.) Academic Press, New York.

Gorski, J. and Nicolette, J. A. (1963). Early estrogen effects on newly synthesized RNA and phospholipid in subcellular fractions of rat uteri. *Arch. Biochem. Biophys.* **103**, 418–423.

Goss, R. J. (1961). Regeneration of vertebrate appendages. *In* "Advances in Morphogenesis", Vol. 1. (Eds Abercrombie, M. and Brachet, J.) Academic Press, New York.

Goss, R. J. (1964). The role of skin in antler regeneration. *In* "Advances in Biology of Skin", Vol. 5. (Ed. Montagna, W.) Pergamon Press, New York.

Goss, R. J. (1964). "Adaptive Growth". Logos Press, London.

Goss, R. J. (1965). Mammalian regeneration and its phylogenetic relationships. *In* "Regeneration in Animals". (Eds Kiortsis, V. and Trampusch, H. A. L.) North-Holland Co., Amsterdam.

Goss, R. J. (1965). The functional demand theory of growth regulation. The stimulation of compensatory renal hyperplasia by functional overload. *In* "Regeneration in Animals and Related Problems". (Eds Kiortsis, V. and Trampusch, H. A. L.) North-Holland Co., Amsterdam.

Goss, R. J. and Rankin, M. (1960). Physiological factors affecting compensatory renal hyperplasia in the rat. *J. Exp. Zool.* **145**, 209–216.

Gowans, J. L. and McGregor, D. D. (1965). The immunological activities of lymphocytes. *Progr. Allergy.* **9**, 1–78.

Granick, S. (1963). The plastids: their morphological and chemical differentiation. *In* "Cytodifferentiation and Macromolecular Synthesis". (Ed. Locke, M.) Academic Press, New York.

Gregg, J. H. (1965). Regulation in the cellular slime molds. *Develop. Biol.* **12**, 377–393.

Griem, M. L. (1966). Use of multiple biopsies for the study of the cell cycle of the mouse hair follicle. *Nature* **210**, 213–214.

Grimstone, A. V. (1961). Fine structure and morphogenesis in Protozoa. *Biol. Rev. Cambridge Phil. Soc.* **36**, 97–150.

Grobstein, C. (1959). Differentiation of vertebrate cells. *In* "The Cell", Vol. I. (Eds Brachet, J. and Mirsky, A. E.) Academic Press, London.

Gros, F. (1964). The genetic code and its translation. *In* "Cellular Control Mechanisms and Cancer". (Eds Emmelot, P. and Mühlbock, O.) Elsevier, Amsterdam.

Grüneberg, H. (1952). "The Genetics of the Mouse". Martinus Nijhoff, The Hague.

Gurdon, J. B. (1962). The transplantation of nuclei between two species of *Xenopus. Develop. Biol.* **5**, 68–83.

Gurdon. J. B. (1963). Nuclear transplantation in Amphibia and the importance of stable nuclear changes in promoting cellular differentiation. *Quart. Rev. Biol.* **38**, 54–78.

Gurdon, J. B. (1964). The transplantation of living cell nuclei. *Advan. Morphogenesis* **4**, 1–43.

Gurdon, J. B. and Uehlinger, V. (1966). "Fertile" intestine nuclei. *Nature* **210**, 1240–1241.

Guttman, H. N. and Wallace, F. G. (1964). Nutrition and physiology of the Trypanosomatidae. *In* "Biochemistry and Physiology of Protozoa". (Ed. Hutner, S. H.) Academic Press, New York.

Haddow, A. (1964). Mechanisms of carcinogenesis. *Brit. Med. Bull.* **20**, 87–90.

Haemmerling, J. (1963). The role of the nucleus in differentiation especially in *Acetabularia. Symp. Soc. Exp. Biol.* **17**, 127–137.

Haemmerling, J. (1963), Nucleo-cytoplasmic interactions in *Acetabularia* and other cells. *Ann. Rev. Plant Physiol.* **14**, 65–92.

Hahn, H. P. von (1966). A model of "regulatory" aging of the cell at the gene level. *J. Gerontol.* **21**, 291–293.

Hall, W. T. and Claus, G. (1963). Ultrastructural studies on the blue-green algal symbiont in *Cyanophora paradoxa* Korschikoff. *J. Cell Biol.* **19**, 551–563.

Halvorson, H. O. (1965). Sequential expression of biochemical events during intracellular differentiation. *Symp. Soc. Gen. Microbiol.* **15**, 343–368.

Hamilton, T. H. (1963). Isotopic studies on estrogen-induced accelerations of ribonucleic acid and protein synthesis. *Proc. Nat. Acad. Sci. U. S.* **49**, 373–379.

Hartwell, L. H. and Magasanik, B. (1963). The molecular basis of histidase induction in *Bacillus subtilis*. *J. Mol. Biol.* **7**, 401–420.

Hauschka, S. D. and Konigsberg, I. R. (1966). The influence of collagen on the development of muscle clones. *Proc. Nat. Acad. Sci. U. S.* **55**, 119–126.

Hayes, W. (1953). Observations on a transmissible agent determining sexual differentiation in *Bact. coli*. *J. Gen. Microbiol.* **8**, 72–88.

Hayes, W. (1953). The mechanism of genetic recombination in *E. coli*. *Cold Spring Harbor Symp. Quant. Biol.* **18**, 75–93.

Hayes, W. (1964). "The Genetics of Bacteria and their Viruses". Blackwell, Oxford.

Hayes, W. (1966). Sex factors and viruses. *Proc. Roy. Soc. (London)*, Ser. B **164**, 230–245.

Hayes, W. (1967). The mechanism of bacterial sexuality. *Endeavour* **26**, 33–38.

Hayflick, L. (1965). The limited *in vitro* lifetime of human diploid cell strains. *Exp. Cell Res.* **37**, 614–636.

Hechter, O. and Halkerston, I. D. K. (1965). Effects of steroid hormones on gene regulation and cell metabolism. *Ann. Rev. Physiol.* **27**, 133–162.

Heilman, F. R. and Kendall, E. C. (1944). The influence of 11-dehydro-17-hydroxycorticosterone on the growth of a malignant tumor in the mouse. *Endocrinology* **34**, 416–420.

Hennen, S. (1963). Chromosomal and embryological analyses of nuclear changes occurring in embryos derived from transfers of nuclei between *Rana pipiens* and *Rana sylvatica*. *Develop. Biol.* **6**, 133–183.

Henson, H. (1946). The theoretical aspect of insect metamorphosis. *Biol. Rev. Cambridge Phil. Soc.* **21**, 1–14.

Herlant-Meewis, H. (1964). Regeneration in annelids. *Advan. Morphogenesis* **4**, 155–215.

Hieger, I. (1959). Theories of carcinogenesis. *In* "Ciba Foundation Symposium on Carcinogenesis". (Eds Wolstenholme, G. E. W. and O'Connor, M.) Churchill, London.

Hieger, I. (1961). "Carcinogenesis". Academic Press, London.

Higgins, G. M. and Woods, K. A. (1950). The influence of cortisone (compound E) upon a lymphoid leukemia induced in AKM mice. *Anat. Record* **106**, 204.

Holliday, M., Bright, N. H., Schulz, D. and Oliver, J. (1961). The renal lesions of electrolyte imbalance. *J. Exp. Med.* **113**, 971–980.

Holtfreter, J. (1933). Nachweis der Induktionsfähigkeit abgetöteter Keimteile. Isolations- und Transplantationsversuche. *Arch. Entwicklungsmech. Organ.* **128**, 584–633.

Holtfreter, J. (1938). Differenzierungspotenzen isolierter Teile der Anurengastrula. *Arch. Entwicklungsmech. Organ.* **138**, 657–738.

Holtfreter, J. (1947). Neural induction in explants which have passed through a sublethal cytolysis. *J. Exp. Zool.* **106**, 197–222.

Homan, J. D. H. and Hondius Boldingh, W. (1965). Personal communication (from N. V. Organon, Oss, The Netherlands).

Hotta, Y. and Stern, H. (1963). Molecular facets of mitotic regulation. *Proc. Nat. Acad. Sci. U. S.* **49**, 648–653; 861–865.

Huang, R. C. and Bonner, J. T. (1962). Histone, a suppressor of chromosomal RNA synthesis. *Proc. Nat. Acad. Sci. U. S.* **48**, 1216–1222.

Hughes, T. E. (1964). Neurosecretion, ecdysis and hypopus formation in the Acaridei. *Proc. Intern. Congr. Acarology*, **1**, 338–342.

Humphreys, T., Penman, S. and Bell, E. (1964). The appearance of stable polysomes during the development of chick down feathers. *Biochem. Biophys. Res. Commun.* **17**, 618–623.

Huskins, C. L. (1947). The subdivision of the chromosomes and their multiplication in non-dividing tissues: possible interpretations in terms of gene structure and gene action. *Am. Naturalist* **81**, 401–434.

Iversen, O. H. (1961). The regulation of cell numbers in epidermis. A cybernetic point of view. *Acta Pathol. Microbiol. Scand., Suppl.* **148**, 91–96.

Iversen, O. H. (1965). Cybernetic aspects of the cancer problem. *Prog. Biocybern.* **2**, 76–110.

Iwai, K. (1964). Histones of rice embryos and of *Chlorella. In* "The Nucleohistones". (Eds Bonner, J. and Ts'o, P. O. P.) Holden-Day, San Francisco.

Jacob, F. (1964). Regulatory devices in the bacterial cell. *In* "Cellular Control Mechanisms and Cancer". (Eds Emmelot, P. and Mühlbock, O.) Elsevier, Amsterdam.

Jacob, F. (1966). Genetics of the bacterial cell. *Science* **152**, 1470–1478.

Jacob, F. and Brenner, S. (1963). Sur la régulation de la synthèse du DNA chez les bactéries: l'hypothèse du réplicon. *Compt. Rend.* **256**, 298.

Jacob, F. and Monod, J. (1961). Genetic regulatory mechanisms in the synthesis of proteins. *J. Mol. Biol.* **3**, 318–356.

Jacob, F. and Monod, J. (1963). Elements of regulating circuits in bacteria. *In* "Biological Organization at the Cellular and Supercellular Level". (Ed. Harris, R. J. C.) Academic Press, London.

Jacob, F. and Monod, J. (1963). Genetic repression, allosteric inhibition, and cellular differentiation. *In* "Cytodifferentiation and Macromolecular Synthesis". (Ed. Locke, M.) Academic Press, New York.

Jacob, F. and Wollman, E. L. (1961). "Sexuality and the Genetics of Bacteria". Academic Press, New York.

Jacob, F. and Wollman, E. L. (1961). Viruses and genes. *Sci. Am.* **204**, 92–107.

Jacob, F., Brenner, S. and Cuzin, F. (1963). On the regulation of DNA replication in bacteria. *Cold Spring Harbor Symp. Quant. Biol.* **28**, 329–348.

Jacob, F., Ullman, A. and Monod, J. (1964). Le promoteur, élément génétique nécessaire à l'expression d'un operon. *Compt. Rend.* **258**, 3125–3128.

Jacobson, L. O. and Doyle, M. (eds) (1962). "Erythropoiesis". Grune and Stratton, New York.

Jaffe, J. J., Fischer, G. A. and Welch, A. D. (1963). Structure-action relationships of corticosteroid compounds as inhibitors of leukemic L5187Y cell reproduction *in vivo* and *in vitro. Biochem. Pharmacol.* **12**, 1081–1090.

Jinks, J. L. (1964). "Extrachromosomal Inheritance". Prentice-Hall, New Jersey.

Johns, E. W. and Butler, J. A. V. (1964). Specificity of the interactions between histones and deoxyribonucleic acid. *Nature* **204**, 853–855.

Jorpes, E. and Mutt, V. (1964). Gastrointestinal hormones. *In* "The Hormones", Vol. 4. (Eds Pincus, G., Thimann, K. V. and Astwood, E. B.) Academic Press, New York.

Kaplan, H. S. (1964). Some possible mechanisms of carcinogenesis. *In* "Cellular Control Mechanisms and Cancer". (Eds Emmelot, P. and Mühlbock, O.) Elsevier, Amsterdam.

Karlson, P. (1963). Chemistry and biochemistry of insect hormones. *Angew. Chem. Intern. Ed. Engl.* **2**, 175–182.

Karlson, P. and Butenandt, A. (1959). Pheromones (ectohormones) in insects. *Ann. Rev. Entomol.* **4**, 39–58.

Karlson, P. and Sekeris, C. E. (1962). Zum Tyrosinstoffwechsel der Insekten. Kontrolle des Tyrosinstoffwechsels durch Ecdyson. *Biochim. Biophys. Acta* **63**, 489–495.

Karlson, P. and Sekeris, C. E. (1966). Biochemical mechanisms of hormone action. *Acta Endocrinol.* **53**, 505–518.

Kassenaar, A., Kouwenhoven, A. and Querido, A. (1962). On the metabolic action of testosterone and related compounds. *Acta Endocrinol.* **39**, 223–233.

Kasten, F. H. and Strasser, F. F. (1966). Nucleic acid synthetic patterns in synchronized mammalian cells. *Nature* **211**, 135–140.

Kimball, R. F. (1964). Physiological genetics of the ciliates. *In* "Biochemistry and Physiology of Protozoa", Vol. 3. (Ed. Hutner, S. H.) Academic Press, New York.

Kiortsis, V. and Trampusch, H. A. L. (eds) (1965). "Regeneration in Animals". North-Holland Co., Amsterdam.

Kit, S. and Barron, E. S. G. (1953). The effect of adrenal cortical hormones on the incorporation of C^{14} into the protein of lymphatic cells. *Endocrinology* **52**, 1–9.

Klein, G. and Klein, E. (1957). The evolution of independence from specific growth stimulation and inhibition in mammalian tumour-cell populations. *Symp. Soc. Exp. Biol.* **11**, 305–328.

Kochakian, C. D. and Harrisson, D. G. (1962). Regulation of nucleic acid synthesis by androgens. *Endocrinology* **70**, 99–108.

Konrad, C. G. (1963). Protein synthesis and RNA synthesis during mitosis in animal cells. *J. Cell Biol.* **19**, 267–277.

Koritz, S. B. and Dorfman, R. I. (1956). Studies on the inhibition by deoxycorticosterone of the *in vitro* incorporation of glycine-1-C^{14} into the proteins of reticulocytes. *Arch. Biochem. Biophys.* **65**, 491–499.

Kornberg, H. L. (1965). The co-ordination of metabolic routes. *Symp. Soc. Gen. Microbiol.* **15**, 8–31.

Kroeger, H. (1963). Experiments on the extranuclear control of gene activity in dipteran polytene chromosomes. *J. Cellular Comp. Physiol.* **62**, suppl. 1, 45–59.

Kroeger, H. and Lezzi, M. (1966). Regulation of gene action in insect development. *Ann. Rev. Entomol.* **11**, 1–22.

Krohn, P. L. (1962). Heterochronic transplantation in the study of ageing. *Proc. Roy. Soc. (London), Ser. B* **157**, 128–147.

Kudo, R. R. (1947). *Pelomyxa carolinensis* Wilson. II. Nuclear division and plasmotomy. *J. Morphol.* **80**, 93–144.

Kudo, R. R. (1951). Observations on *Pelomyxa illinoisensis*. *J. Morphol.* **88**, 145–173.

Kudo, R. R. (1966). "Protozoology". C. C. Thomas, Springfield.

Lajtha, L. G. and Oliver, R. (1960) Studies on the kinetics of erythropoiesis. *In* "Ciba Foundation Symposium on Haemopoiesis". (Eds Wolstenholme, G. E. W. and O'Connor, M.) Churchill, London.

Lajtha, L. G., Gilbert, C. W., Proteous, D. D. and Alexanian, A. (1964). Kinetics of a bone-marrow stem-cell population. *Ann. N. Y. Acad. Sci.* **113**, 742–752.

Lamerton, L. F. (1964). Radiation carcinogenesis. *Brit. Med. Bull.* **20**, 134–138.

Lamerton, L. F. (1966). Cell proliferation under continuous radiation. *Radiation Res.* **27**, 119–138.

Lang, N. and Sekeris, C. E. (1964). Stimulation of RNA-polymerase activity in rat liver by cortisol. *Life Sci.* **3**, 391–394.

Lanni, F. (1964). The biological coding problem. *Advan. Genet.* **12**, 1–141.

Lash, J. W., Holtzer, H. and Swift, H. (1957). Regeneration of mature skeletal muscle. *Anat. Record* **128**, 679–697.

Law, L. W. (1947). Effect of gonadectomy and adrenalectomy on the appearance and incidence of spontaneous lymphoid leukemia in C58 mice. *J. Nat. Cancer Inst.* **8**, 157–159.

Laws, J. O. (1959). Tissue regeneration and tumour development. *Brit. J. Cancer* **13**, 669–674.

Laws, J. O. (1960). Some biological aspects of carcinogenesis in the liver. *Acta Unio Intern. Contra Cancrum* **16**, 87–90.

Laws, J. O. (1966). Escape from control. *Sci. J.* **2**, 98.

Leblond, C. P. and Sainte-Marie, G. (1960). Models for lymphocyte and plasmocyte production. *In* "Ciba Foundation Symposium on Haemopoiesis". (Eds Wolstenholme, G. E. W. and O'Connor, M.) Churchill, London.

Leblond, C. P. and Walker, B. E. (1956). Renewal of cell populations. *Physiol. Rev.* **36**, 255–276.

Lederberg, J. and Tatum, E. L. (1946). Gene recombination in *E. coli. Nature* **158**, 558.

Lehman, I. R., Bessman, M. J., Simms, E. S. and Kornberg, A. (1958). Enzymatic synthesis of DNA. *J. Biol. Chem.* **233**, 163–170.

Leong, G. F., Grisham, J. W., Hole, B. V. and Albright, M. L. (1964). Effect of partial hepatectomy on DNA synthesis and mitosis in heterotopic partial autografts of rat liver. *Cancer Res.* **24**, 1496–1501.

Lesher, S., Fry, R. J. M. and Kohn, H. I. (1961). Influence of age on transit time of cells of mouse intestinal epithelium. *Lab. Invest.* **10**, 291–300.

Lewis, E. B. (1964). Genetic control and regulation of developmental pathways. *In* "The Role of Chromosomes in Development". (Ed. Locke, M.) Academic Press, New York.

Leyden, C. E. D. F. van (1916). Some observations on periodic nuclear division in the cat. *Proc. Koninkl. Ned. Akad. Wetenschap.*, **19**, 38–44.

Leyden, C. E. D. F. van (1926). Day and night period in nuclear divisions. *Proc. Koninkl. Ned. Akad. Wetenschap.* **29**, 979–988.

Liao, S. and Williams-Ashman, H. G. (1962). An effect of testosterone on amino acid incorporation by prostatic ribonucleoprotein particles. *Proc. Nat. Acad. Sci. U. S.* **48**, 1956–1964.

Lieberman, I. and Ove, P. (1962). Deoxyribonucleic acid synthesis and its inhibition in mammalian cells cultured from the animal. *J. Biol. Chem.* **237**, 1634–1642.

Lopashov, G. (1936). Eye inducing substances. *Inst. Exp. Biol.*, Moscow.

Lotspeich, W. D. (1965). Renal hypertrophy in metabolic acidosis and its relation to ammonia excretion. *Am. J. Physiol.* **208**, 1135–1142.

Lwoff, A. (1950). "Problems of Morphogenesis in Ciliates". John Wiley, New York.

Lwoff, A. (1966). Interaction among virus, cell, and organism. *Science* **152**, 1216–1219.

Lynch, H. T., Shaw, M. W., Magnuson, C. W., Larsen, A. L. and Krush, A. J. (1966). Hereditary factors in cancer: study of two large Midwestern kindreds. *Arch. Internal Med.* **117**, 206–212.

Maaløe, O. (1961). The control of normal DNA replication in bacteria. *Cold Spring Harbor Symp. Quant. Biol.* **26**, 45–52.

Maaløe, O. (1963). Role of protein synthesis in the DNA replication cycle in bacteria. *J. Cellular Comp. Physiol.* **62**, suppl. 1, 31–44.

MacDonald, R. A. (1961). "Lifespan" of liver cells. *Arch. Internal Med.* **107**, 335–343.

Mangold, O. (1961). Grundzüge der Entwicklungsphysiologie der Wirbeltiere mit besonderer Berücksichtigung der Missbildungen auf Grund experimenteller Arbeiten an Urodelen. *Acta Genet. Med. Gemell.* **10**, 1–49.

Markert, C. L. (1965). Mechanisms of cellular differentiation. *In* "Ideas in Modern Biology". (Ed. Moore, J. A.) Doubleday, New York.

Marsh, J., Miller, B. E. and Lamson, B. G. (1959). Effect of repeated brief stress on the growth of Ehrlich carcinoma in the mouse. *J. Nat. Cancer Inst.* **22**, 971–977.

Maynard Smith, J. (1962). The causes of ageing. *Proc. Roy. Soc. (London), Ser. B* **157**, 115–127.

Mazia, D. (1961). Mitosis and the physiology of cell division. *In* "The Cell", Vol. 3. (Eds Brachet, J. and Mirsky, A. E.) Academic Press, New York.

Mazia, D. (1961). How cells divide. *Sci. Am.* **205**, 100–112.

Mazia, D. (1963). Synthetic activities leading to mitosis. *J. Cellular Comp. Physiol.* **62**, suppl. 1, 123–140.

Mazia, D. (1965). The partitioning of genomes. *Symp. Soc. Gen. Microbiol.* **15**, 379–394.

Mechelke, F. (1953). Reversible Strukturmodifikationen der Speicheldrüsen-chromosome von *Acricotopus lucidus*. *Chromosoma* **5**, 511–543.

Medawar, P. B. (1963). Definition of the immunologically competent cell. *In* "The Immunologically Competent Cell". (Eds Wolstenholme, G. E. W. and Knight, J.) Churchill, London.

Meerson, F. Z. (1965). Intensity of function of structures of the differentiated cell as a determinant of activity of its genetic apparatus. *Nature* **206**, 483–484.

Mercer, E. H. (1962). The cancer cell. *Brit. Med. Bull.* **18**, 187–192.

Metcalf, D. (1960). Adrenal cortical function in high- and low-leukemia strains of mice. *Cancer Res.* **20**, 1347–1353.

Miksche, J. P. (ed.) (1964). Meristems and differentiation. *Brookhaven Symp. Biol.* **16**, 1–240.

Mohn, M. P. (1958). The effects of different hormonal states on the growth of hair in rats. *In* "The Biology of Hair Growth". (Eds Montagna, W. and Ellis, R. A.) Academic Press, London.

Molomut, N., Lazere, F. and Smith, L. W. (1963). Effect of audiogenic stress upon methylcholanthrene-induced carcinogenesis in mice. *Cancer Res.* **23**, 1097–1101.

Monod, J. and Jacob, F. (1961). General conclusions: teleonomic mechanisms in cellular metabolism, growth, and differentiation. *Cold Spring Harbor Symp. Quant. Biol.* **26**, 389–401.

Monod, J., Changeux, J. P. and Jacob, F. (1963). Allosteric proteins and cellular control systems. *J. Mol. Biol.* **6**, 306–329.

Montagna, W. (1962). "The Structure and Function of Skin". Academic Press, New York.

Montagna, W. and Hamilton, J. B. (1949). Mitotic activity in the epidermis of the rabbit stimulated with local applications of testosterone propionate. *J. Exp. Zool.* **110**, 379–396.

Montagna, W. and Kenyon, P. (1949). Growth potentials and mitotic division in the sebaceous glands of the rabbit. *Anat. Record* **103**, 365–380.

Moore, J. A. (1958). The transfer of haploid nuclei between *Rana pipiens* and *Rana sylvatica*. *Exp. Cell Res.*, suppl. 6, 179–191.

Moore, J. A. (1960). Serial back-transfers of nuclei in experiments involving two species of frogs. *Develop. Biol.* **2**, 535–550.

Moore, J. A. (1962). Nuclear transplantation and problems of specificity in developing embryos. *J. Cellular Comp. Physiol.* **60**, suppl. 1, 19–34.

Nanney, D. L. (1953). Nucleo-cytoplasmic interaction during conjugation in *Tetrahymena*. *Biol. Bull.* **105**, 133–148.

Nanney, D. L. (1964). Macronuclear differentiation and subnuclear assortment in ciliates. *In* "The Role of Chromosomes in Development". (Ed. Locke, M.) Academic Press, New York.

Nanney, D. L. and Rudzinska, M. A. (1960). Protozoa. *In* "The Cell", Vol. 4. (Eds Brachet, J. and Mirsky, A. E.) Academic Press, New York.

Needham, A. E. (1960). Regeneration and growth. *In* "Fundamental Aspects of Normal and Malignant Growth". (Ed. Nowinski, W. W.) Elsevier, London.

Needham. J. (1942). "Biochemistry and Morphogenesis". Cambridge University Press.

Nirenberg, M. W. (1963). The genetic code. *Sci. Am.* **208**, 80–93.

Nobili, R. (1961). L'azioni del gene am sull'apparato nucleare di *Paramecium aurelia* durante la riproduzione vegetativa e sessuale in relazione all'età del clone ed alla temperatura di allevamento degli animali. *Caryologia* **14**, 43–58.

Nossal, G. J. V. (1965). The mechanism of action of antigen. *Australasian Ann. Med.* **14**, 321–328.

Nossal, G. J. V. and Mäkelä, O. (1962). Autoradiographic studies on the immune response. *J. Exp. Med.* **115**, 209–230.

Nowell, P. C. (1960). Phytohemagglutinin: an initiator of mitosis in cultures of normal human leukocytes. *Cancer Res.* **20**, 462–466.

Oordt, G. J. van (1963). Male gonadal hormones. *In* "Comparative Endocrinology", Vol. 1. (Eds Euler, U.S. von and Heller, H.) Academic Press, London.

Oparin, A. I. (1957). "The Origin of Life on the Earth". (3rd Edition.) Academic Press, New York.

Oparin, A. I. (1962). "Life, its Nature, Origin and Development". Academic Press, New York.

Osgood, E. E. (1957). A unifying concept of the etiology of the leukemias, lymphomas, and cancers. *J. Nat. Cancer Inst.* **18**, 155–166.

Osgood, E. E. (1959). Regulation of cell proliferation. *In* "The Kinetics of Cellular Proliferation". (Ed. Stohlman, F.) Grune and Stratton, New York.

Otsuka, H. and Terayama, H. (1966). Inhibition of DNA synthesis in ascites hepatoma cells by normal liver extract. *Biochim. Biophys. Acta* **123**, 274–285.

Paschkis, K. E. (1958). Growth-promoting factors in tissues: a review. *Cancer Res.* **18**, 981–991.

Pasteels, J. (1940). Recherches sur les facteurs initiaux de la morphogénèse chez les Amphibiens Anoures. III. *Arch. Biol.* (*Liège*) **51**, 103–150.

Pasteels, J. (1941). Recherches sur les facteurs initiaux de la morphogénèse chez les Amphibiens Anoures. V. *Arch. Biol.* (*Liège*) **52**, 321–339.

Pasteels, J. (1948). Recherches sur le cycle germinal chez l'*Ascaris*. *Arch. Biol.* (*Liège*) **59**, 405–446.

Pearson, A. E. G. (1959). The effects of reduction in food intake on growth and differentiation of a squamous-celled carcinoma in mice. *Brit. J. Cancer* **11**, 470–474.

Peckham, B. and Kiekhofer, W. (1962). Cellular behavior in the vaginal epithelium of estrogen-treated rats. *Am. J. Obstet. Gynecol.* **83**, 1021–1027.

Pelling, G. (1959). Chromosomal synthesis of ribonucleic acid as shown by incorporation of uridine labelled with tritium. *Nature* **184**, 655–656.

Pietsch, P. (1961). The effects of colchicine on regeneration of mouse skeletal muscle. *Anat. Record* **139**, 167–172.

Pitelka, D. R. and Child, F. M. (1964). The locomotor apparatus of ciliates and flagellates: relations between structure and function. *In* "Biochemistry and Physiology of Protozoa". (Ed. Hutner, S. H.) Academic Press, New York.

Prehn, R. T. (1960). Tumor-specific immunity to transplanted dibenz(a,h)-anthracene-induced sarcomas. *Cancer Res.* **20**, 1614–1617.

Prescott, D. M. (1964). Turnover of chromosomal and nuclear proteins. *In* "The Nucleohistones". (Eds Bonner, J. and Ts'o, P. O. P.) Holden-Day, San Francisco.

Prescott, D. M. and Bender, M. A. (1963). Synthesis and behavior of nuclear proteins during the cell life cycle. *J. Cellular Comp. Physiol.* **62**, 175–194.

Price, D. and Williams-Ashman, H. G. (1961). The accessory reproductive glands of mammals. *In* "Sex and Internal Secretions". (Ed. Young, W. C.) Williams and Wilkins, Baltimore.

Prop, F. J. A. (1965). Personal communication.

Pullinger, B. D. (1945). An experimental approach to the problem of trauma and tumours. *J. Pathol. Bacteriol.* **57**, 467–477.

Pullinger, B. D. (1945). A measure of the stimulating effect of simple injury combined with carcinogenic chemicals on tumour formation in mice. *J. Pathol. Bacteriol.* **57**, 477–481.

Ramakrishnan, T. and Adelberg, E. A. (1965). Regulatory mechanisms in the biosynthesis of isoleucine and valine. *J. Bacteriol.* **89**, 661–664.

Randle, P. J. (1964). Insulin. *In* "The Hormones", Vol. 4. (Eds Pincus, G., Thimann, K. V. and Astwood, E. B.) Academic Press, New York.

Rashkis, H. A. (1952). Systemic stress as an inhibitor of experimental tumors in Swiss mice. *Science* **116**, 169–171.

Raven, C. P. (1958). "Morphogenesis: the Analysis of Molluscan Development". Pergamon Press, London.

Raven, C. P. (1963). Differentiation in mollusc eggs. *Symp. Soc. Exp. Biol.* **17**, 274–284.

Richardson, C. C., Schildkraut, C. L. and Kornberg, A. (1963). Studies on the replication of DNA by DNA polymerases. *Cold Spring Harbor Symp. Quant. Biol.* **28**, 9–18.

Roe, F. J. C. and Ambrose, E. J. (1966). Future strategy. *In* "The Biology of Cancer". (Eds Ambrose, E. J. and Roe, F. J. C.) Van Nostrand, London.

Roels, F. (1964). L'hypertrophie compensatrice du rein. *Biol. Jaarboek Konink. Natuurw. Genoot. Dodonaea Gent*, **32**, 258–291.

Roels, F. (1965). Inhibition de l'activité mitotique au cours de l'hypertrophie compensatrice du rein par injection d'homogénats rénaux. *Compt. Rend. Soc. Biol.* **159**, 495.

Rosenberg, B. H. and Cavalieri, L. F. (1965). Template deoxyribonucleic acid and the control of replication. *Nature* **206**, 999–1001.

Rous, P. (1959). Surmise and fact on the nature of cancer. *Nature* **183**, 1357–1361.

Rous, P. and Kidd, J. G. (1941). Conditional neoplasms and sub-threshold neoplastic states. *J. Exp. Med.* **73**, 365–389.

Rowley, D. B. and Newcomb, H. R. (1964). Radiosensitivity of several dehydrogenases and transaminases during sporogenesis of *Bacillus subtilis*. *J. Bacteriol.* **87**, 701–709.

Rusch, H. P. and Kline, B. E. (1944). The effect of exercise on the growth of a mouse tumor. *Cancer Res.* **4**, 116–118.

Rusch, H. P., Johnson, R. O. and Kline, B. E. (1945). The relation of caloric intake and of blood sugar to sarcogenesis in mice. *Cancer Res.* **5**, 705–712.

Rusch, H. P., Kline, B. E. and Baumann, C. A. (1945). The influence of caloric restriction and of dietary fat on tumor formation with ultraviolet radiation. *Cancer Res.* **5**, 431–435.

Rusch, H. P., Braun, R., Daniel, J. W. Mittermayer, C. and Sachsenmaier, W. (1964). The role of DNA and RNA in mitosis and differentiation in *Physarum polycephalum*. *In* "Cellular Control Mechanisms and Cancer". (Eds Emmelot, P. and Mühlbock, O.) Elsevier, Amsterdam.

Ryter, A. and Jacob, F. (1963). Étude au microscope électronique des relations entre mésosomes et noyaux chez *Bacillus subtilis*. *Compt. Rend.* **257**, 3060.

Rytömaa, T. (1967). Personal communication.

Rytömaa, T. and Kiviniemi, K. (1967). Regulation system of blood cell production. *In* "Control of Cellular Growth in Adult Organisms". (Eds Teir, H. and Rytömaa, T.) Academic Press, London.

Saetren, H. (1956). A principle of auto-regulation of growth. Production of organ specific mitose-inhibitors in kidney and liver. *Exp. Cell Res.* **11**, 229–232.

Sager, R. (1964). Studies in cell heredity with *Chlamydomonas*. *In* "Biochemistry and Physiology of Protozoa". (Ed. Hutner, S. H.) Academic Press, New York

Sager, R. (1965). Genes outside the chromosomes. *Sci. Am.* **212**, 70–79.

Sager, R. (1966). Mendelian and non-Mendelian heredity: a reappraisal. *Proc. Roy. Soc. (London), Ser. B* **164**, 290–297.

Saunders, J. W., Gasseling, M. T. and Saunders, L. C. (1962). Cellular death in morphogenesis of the avian wing. *Develop. Biol.* **5**, 147–178.

Saxén, L. and Toivonen, S. (1962). "Primary Embryonic Induction". Academic Press, London.

Sayers, G. (1950). The adrenal cortex and homeostasis. *Physiol. Rev.* **30**, 241–320.

Schaeffer, P., Ionesco, H., Ryter, A. and Balassa, G. (1964). La sporulation de *Bacillus subtilis*; étude génétique et physiologique. *Colloq. Intern. Centre Nat. Rech. Sci. (Paris)* **124**, 553–563.

Schäfer, E. A. (1916). "The Endocrine Organs". Longman Green, London.

Schatten, W. E. (1958). An experimental study of post-operative tumor metastases. I. Growth of pulmonary metastases following total removal of primary leg tumor. *Cancer* **11**, 455–459.

Scherbaum, O. H. and Loeffer, J. B. (1964). Environmentally induced growth oscillations in Protozoa. *In* "Biochemistry and Physiology of Protozoa". (Ed. Hutner, S. H.) Academic Press, New York.

Schiff, J. A. and Epstein, H. T. (1965). The continuity of the chloroplast in *Euglena. In* "Reproduction: Molecular, Subcellular, and Cellular". (Ed. Locke, M.) Academic Press, New York.

Schlesinger, S. and Magasanik, B. (1964). Effect of α-methylhistidine on the control of histidine synthesis. *J. Mol. Biol.* **9**, 670–682.

Schneider, D. (1966). Chemical sense communication in insects. *Symp. Soc. Exp. Biol.* **20**, 273–297.

Schultz, J. (1952). Interrelations between nucleus and cytoplasm: problems at the biological level. *Exp. Cell Res., Suppl.* **2**, 17–43.

Scott, E. J. van and Ekel, T. M. (1963). Kinetics of hyperplasia in psoriasis. *Arch. Dermatol.* **88**, 373–381.

Scott, R. B. and Bell, E. (1964). Protein synthesis during development: control through messenger RNA. *Science* **145**, 711–713.

Seaman, G. R. (1960). Large-scale isolation of kinetosomes from the ciliated protozoan *Tetrahymena pyriformis. Exp. Cell Res*, **21**, 292–302.

Seaman, G. R. (1962). Protein synthesis by kinetosomes isolated from the protozoan *Tetrahymena. Biochim. Biophys. Acta* **55**, 889–899.

Sekeris, C. E. (1965). Action of ecdysone on RNA and protein in metabolism in the blowfly, *Calliphora erythrocephala. In* "Mechanisms of Hormone Action". (Ed. Karlson, P.) Academic Press, London.

Sekeris, C. E. and Lang, N. (1964). Induction of dopa-decarboxylase activity by insect messenger RNA in an *in vitro* amino acid incorporating system from rat liver. *Life Sci.* **3**, 625–632.

Setälä, K. (1965). Differences in pharmacodynamic response to colchicine between benign and malignant epidermal hyperplasias. *Acta Radiol., Suppl.* **237**, 1–89.

Shaffer, B. M. (1961). The cells founding aggregation centres in the slime mould, *Polysphondylium violaceum. J. Exp. Biol.* **38**, 833–849.

Shaffer, B. M. (1963). Inhibition by existing aggregates of founder differentiation in the cellular slime mould *Polysphondylium violaceum. Exp. Cell Res.* **31**, 432–435.

Shaffer, B. M. (1964). The Acrasina. *Advan. Morphogenesis* **3**, 301–322.

Singer, M. (1954). Induction of regeneration of the forelimb of the postmetamorphic frog by augmentation of the nerve supply. *J. Exp. Zool.* **126**, 419–472.

Singer, M. (1958). The regeneration of body parts. *Sci. Am.* **199**, 79–88.

Singer, M. (1960). Nervous mechanisms in the regeneration of body parts in vertebrates. *In* "Developing Cell Systems and their Control". (Ed. Rudnick, D.) Ronald Press, New York.

Sirlin, J. L. (1960). Cell sites of RNA and protein synthesis in the salivary gland of *Smittia* (Chironomidae). *Exp. Cell Res.* **19**, 177–180.

Smithers, D. W. (1967). The neoplastic response. *In* "Control of Cellular Growth in Adult Organisms". (Eds Teir, H. and Rytömaa, T.) Academic Press, London.

Sneath, P. H. A. (1962). Longevity of micro-organisms. *Nature* **195**, 643–646.

Sonneborn, T. M. (1954). Gene-controlled, aberrant nuclear behavior in *Paramecium aurelia. Microbiol. Genet. Bull.* **11**, 24–25.

Sonneborn, T. M. (1954). Is gene K active in the micronucleus of *Paramecium aurelia? Microbiol. Genet. Bull.* **11**, 25–26.

Sonneborn, T. M. (1954). Patterns of nucleocytoplasmic integration in *Paramecium. Caryologia, Suppl.* **6**, 307–325.

Sonneborn, T. M. (1959). Kappa and related particles in *Paramecium. Advan. Virus Res.* **6**, 229–356.

Sonneborn, T. M. (1960). The gene and cell differentiation. *Proc. Nat. Acad. Sci. U. S.* **46**, 149–165.

Sonneborn, T. M. (1963). Does preformed cell structure play an essential role in cell heredity? *In* "The Nature of Biological Diversity". (Ed. Allen, J. M.) McGraw-Hill, New York.

Spemann, H. (1901). Entwicklungsphysiologische Studien am *Triton*-Ei. *Arch. Entwicklungsmech. Organ.* **12**, 224–264.

Spemann, H. (1912). Zur Entwicklung des Wirbeltierauges. *Zool. Jahresber.* **32**, 1–98.

Spemann, H. (1938). "Embryonic Development and Induction". Yale University Press, New Haven, Connecticut.

Srb, A. M. (1963). Extrachromosomal factors in the genetic differentiation of *Neurospora. Symp. Soc. Exp. Biol.* **17**, 175–187.

Srb, A. M., Owen, R. D. and Edgar, R. S. (1965). "General Genetics". Freeman and Company, San Francisco.

Srinivasan, V. R. and Halvorson, H. O. (1963). "Endogenous factor" in sporogenesis in bacteria. *Nature* **197**, 100–101.

Starling, E. H. (1906). Die chemische Koordination der Körpertätigkeiten. *Verhand. Gesellsch. dt. Naturforsch. Ärzte (Stuttgart)* pt 1, 246–260.

Starling, E. H. (1914). Discussion on the therapeutic value of hormones. *Proc. Roy. Soc. Med. (Ther. Pharm. Sect.)* **7**, 29–31.

Steinert, G., Firket, H. and Steinert, M. (1958). Synthèse d'acide désoxyribonucleique dans le corps parabasal de *Trypanosoma mega. Exp. Cell Res.* **15**, 632–634.

Stent, G. S. (1964). The operon: on its third anniversary. *Science* **144**, 816–820.

Steward, F. C. (1963). The control of growth in plant cells. *Sci. Am.* **209**, 104–113.

Steward, F. C. and Ram, H. Y. M. (1961). Determining factors in cell growth. *In* "Advances in Morphogenesis", Vol. 1. (Eds Abercrombie, M. and Brachet, J.) Academic Press, New York.

Steward, F. C., Mapes, M. O. and Kent, A. E. (1963). Carrot plants from cultured cells: new evidence for the totipotency of somatic cells. *Am. J. Botany* **50**, 618.

Stich, H. and Plaut, W. (1958). The effect of ribonuclease on protein synthesis in nucleated and enucleated fragments of *Acetabularia. J. Biophys. Biochem. Cytol.* **4**, 119–121.

Stockdale, F. E. and Topper, Y. J. (1966). The role of DNA synthesis and mitosis in hormone-dependent differentiation. *Proc. Nat. Acad. Sci. U. S.* **56**, 1283–1289.

Stoerk, H. C. (1950). Growth retardation of lymphosarcoma implants in pyridoxine-deficient rats by testosterone and cortisone. *Proc. Soc. Exp. Biol. Med.* **74**, 798–800.

Stohlman, F. (1959). Observations on the physiology of erythropoietin and its role in the regulation of red cell production. *Ann. N. Y. Acad. Sci.* **77**, 710–724.

Strehler, B. L. (1962). "Time, Cells, and Aging". Academic Press, New York.

Stretton, A. O. W. (1965). The genetic code. *Brit. Med. Bull.* **21**, 229–235.

Sugiura, K., Stock, C. C., Dobriner, K. and Rhoads, C. P. (1950). The effect of cortisone and other steroids on experimental tumors. *Cancer Res.* **10**, 244–245.

Takenchi, I. (1963). Immunochemical and immunohistochemical studies on the development of the cellular slime mould *Dictyostelium mucoroides. Develop. Biol.* **8**, 1–26.

Talwar, G. S., Segal, S. J., Evans, A. and Davidson, O. W. (1964). The binding of estradiol in the uterus: a mechanism for derepression of RNA synthesis. *Proc. Nat. Acad. Sci. U. S.* **52**, 1059–1066.

Talwar, G. S., Modi, S. and Rao, K. N. (1965). DNA dependent synthesis of RNA is not implicated in growth response of chick comb to androgens. *Science* **150**, 1315–1316.

Tamiya, H. (1963). Cell differentiation in *Chlorella*. *Symp. Soc. Exp. Biol.* **17**, 188–214.

Tannenbaum, A. (1940). The initiation and growth of tumors. I. Effects of under-feeding. *Am. J. Cancer* **38**, 335–350.

Tannenbaum, A. (1942). The genesis and growth of tumors. II. Effects of caloric restriction *per se*. *Cancer Res.* **2**, 460–467.

Tannenbaum, A. (1947). Effects of varying caloric intake upon tumor incidence and tumor growth. *Ann. N. Y. Acad. Sci.* **49**, 5–18.

Tannenbaum, A. (1947). The role of nutrition in the origin and growth of tumors. *In* "Approaches to Tumor Chemotherapy". (Ed. Moulton, F. R.) *Am. Assoc. Advan. Sci.*, Washington.

Tartar, V. (1961). "The Biology of *Stentor*". Pergamon Press, New York.

Teir, H., Rytömaa, T., Cederberg, A. and Kiviniemi, K. (1963). Studies on the elimination of granulocytes in the intestinal tract in rat. *Acta Pathol. Microbiol. Scand.* **59**, 311–324.

Thomas, L. (1959). Mechanisms involved in tissue damage by the endotoxias of gramnegative bacteria. *In* "Cellular and Humoral Aspects of the Hypersensitive States". (Ed. Lawrence, H. S.) Cassell, London.

Todorov, I. N., Blok, L. N. and Vasilchenko, V. N. (1966). The adrenocorticotropic factor synthesised in the acellular system from *Escherichia coli* B. under the influence of ribonucleic acid from bovine pituitary. *Proc. Acad. Sci. USSR, Chem. Sect. (English Transl.)* **167**, 461–463.

Tomkins, G. M. and Maxwell, E. S. (1963). Some aspects of steroid hormone action. *Ann. Rev. Biochem.* **32**, 677–708.

Trainin, N. (1963). Adrenal imbalance in mouse skin carcinogenesis. *Cancer Res.* **23**, 415–419.

Trainin, N., Kaye, A. M. and Berenblum, I. (1964). Influence of mutagens on the initiation of skin carcinogenesis. *Biochem. Pharmacol.* **13**, 263–267.

Trotter, N. L. (1961). The effect of partial hepatectomy on subcutaneously transplanted hepatomas in mice. *Cancer Res.* **21**, 778–782.

Tsanev, R. (1959). Reactive tissue changes following periodic stoppage of blood circulation. *Compt. Rend. Acad. Bulgare Sci.* **12**, 565–568.

Tsanev, R. (1962). Injury-induced changes in nucleic acid content of the epidermis. *Bull. Cent. Biochem. Lab. (Sofia)* **1**, 7–13.

Tsanev, R. (1964). Role of nucleic acids in the wound healing process. *Symp. Biol. Hungarica* **3**, 55–73.

Tsanev. R. and Markov, G. G. (1964). Early changes in liver nucleic acids after partial hepatectomy. *Folia Histochem. Cytochem.* **2**, 233–242.

Ui, H. and Mueller, G. C. (1963). The role of RNA synthesis in early estrogen action. *Proc. Nat. Acad. Sci. U.S.* **50**, 256–260.

Upton, A. C. and Furth, J. (1953). Inhibition by cortisone of the development of spontaneous leukemia and its induction by radiation. *Proc. Am. Ass. Cancer Res.* **1**, 57.

Ursprung, H. (1965). Genes and development. *In* "Organogenesis". (Eds DeHaan, R. L. and Ursprung, H.) Holt, Rinehart, and Winston, New York.

Valentine, R. C. (1966). Sexual differentiation in *E. coli. Biochem. Biophys. Res. Commun.* **22**, 156–162.

Vince, D. (1964). Photomorphogenesis in plant stems. *Biol. Rev. Cambridge Phil. Soc.* **39**, 506–536.

Walford, R. L. (1962). Auto-immunity and aging. *J. Gerontol.* **17**, 281–285.

Wallace, E. W. Wallace, H. and Mills, C. A. (1944). Influence of environmental temperature upon the incidence and course of spontaneous tumors in C_3H mice. *Cancer Res.* **4**, 279–281.

Wallace, E. W., Wallace, H. and Mills, C. A. (1945). Influence on environmental temperature upon the incidence and course of spontaneous tumors in spayed C_3H mice. *Cancer Res.* **5**, 47–48.

Weaver, N. (1966). Physiology of caste determination. *Ann. Rev. Entomol.* **11**, 79–102.

Weber, G. and Convery, H. J. H. (1966). Insulin: inducer of glucose-6-phosphate dehydrogenase. *Life Sci.* **5**, 1139–1146.

Weinbren, K. (1959). Regeneration of the liver. *Gastroenterology* **37**, 657–668.

Weiner, N. (1964). The catecholamines. *In* "The Hormones", Vol. 4. (Eds Pincus, G., Thimann, K. V. and Astwood, E. B.) Academic Press, New York.

Weinstein, G. D. and Scott, E. van (1965). Autoradiographic analysis of turnover times of normal and psoriatic epidermis. *J. Invest. Dermatol.* **45**, 257–262.

Weiss, P. (1955). Specificity in growth control. *In* "Biological Specificity and Growth". (Ed. Butler, E. G.) Princeton University Press.

Wessells, N. K. (1964). DNA synthesis, mitosis, and differentiation in pancreatic acinar cells *in vitro J. Cell Biol.* **20**, 415–433.

Wessells, N. K. (1964). Tissue interactions and cytodifferentiation. *J. Exp. Zool.* **157**, 139–152.

White, A., Blecher, M. and Jedeikin, L. (1961). Mechanism of action of adrenal cortical hormone. *In* "Mechanism of Action of Steroid Hormones". (Eds Villee, C. A. and Engel, L. L.) Pergamon Press, New York.

White, F. R. (1961). The relationship between underfeeding and tumor formation, transplantation, and growth in rats and mice. *Cancer Res.* **21**, 281–290.

Wiese, L. and Jones, R. (1963). Studies on gamete copulation in heterothallic chlamydomonads. *J. Cellular Comp. Physiol.* **61**, 265–274.

Wigglesworth, V. B. (1954). "The Physiology of Insect Metamorphosis". Cambridge University Press.

Wigglesworth, V. B. (1959). Metamorphosis and differentiation. *Sci. Am.* **200**, 100–110.

Wigglesworth, V. B. (1964). Homeostasis in insect growth. *Symp. Soc. Exp. Biol.* **18**, 265–281.

Wigglesworth, V. B. (1965). "The Principles of Insect Physiology". (6th Edition.) Methuen, London.

Wigglesworth, V. B. (1965). The juvenile hormone. *Nature* **208**, 522–524.

Wilkie, D. (1964). "The Cytoplasm in Heredity". Methuen, London.

Willmer, E. N. (1963). Differentiation in *Naegleria. Symp. Soc. Exp. Biol.* **17**, 215–233.

Wood, W. B. and Berg, P. (1962). The effect of enzymatically synthesized ribonucleic acid in amino acid incorporation by a soluble protein-ribosome system from *Escherichia coli. Proc. Nat. Acad. Sci. U. S.* **48**, 94–104.

Wooley, G. W. and Peters, B. A. (1953). Prolongation of life in high-leukemia AKR mice by cortisone. *Proc. Soc. Exp. Biol. Med.* **82**, 286–287.

Wright, B. E. (1963). Endogenous substrate control in biochemical differentiation, *Bacteriol. Rev.* **27**, 273–281.

Wright, B. E. (1964). Biochemistry of Acrasiales. *In* "Biochemistry and Physiology of Protozoa", Vol. 3. (Ed. Hutner, S. H.) Academic Press, New York.

Wright, B. E. (1964). The biochemistry of morphogenesis. *In* "Comparative Biochemistry". (Eds Florkin, M. and Mason, H. S.) Academic Press, London.

Wright, B. E. (1966). Multiple causes and controls in differentiation. *Science* **153**, 830–836.

Yaffe, D. and Feldman, M. (1964). The effect of actinomycin D on heart and thigh muscle cells grown *in vitro*. *Develop. Biol.* **9**, 347–366.

Yntema, C. L. (1959). Regeneration in sparsely innervated and aneurogenic forelimbs of *Amblystoma* larvae. *J. Exp. Zool.* **140**, 101–123.

Yoffey, J. M. (1960). The lymphomyeloid complex. *In* "Haemopoiesis". (Eds Wolstenholme, G. E. W. and O'Connor, M.) Churchill, London.

Yoffey, J. M. (1964). The lymphocyte. *Ann. Rev. Med.* **15**, 125–148.

Yoffey, J. M. (1966). Lymphocytes—the fourth circulation. *Discovery* **27**, 24–29.

Subject Index